Salters-Nuffield Advanced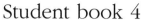
A2 level
Student book 4
Trial Edition

This trial edition may contain typographical errors and other small faults because it is a first proof of the material. We hope this will not detract from reading it. Errors will be rectified in the final published version.

Heinemann Educational Publishers
Halley Court, Jordan Hill, Oxford, OX2 8EJ
Part of Harcourt Education
Heinemann is a registered trademark of Harcourt Education Limited

© Science Education Group, University of York, 2003

Copyright notice

All rights reserved. No part of this publication may be reproduced in any form or by any means (including photocopying or storing it in any medium by electronic means and whether or not transiently or incidentally to some other use of this publication) without the written permission of the copyright owner, except in accordance with the provisions of the Copyright, Designs and Patents Act 1988 or under the terms of a licence issued by the Copyright Licensing Agency Ltd, 90 Tottenham Court Road, London W1P 4LP. Applications for the copyright owner's written permission should be addressed to the publisher.

First published 2003

ISBN 0 435 62837 2

07 06 05 04 03

10 9 8 7 6 5 4 3 2 1

Designed and typeset by J&L Composition, Filey

Printed and bound in Great Britain by The Bath Press Ltd, Bath

Original Illustrations © Harcourt Education Limited, 2003

Acknowledgements

The authors and publishers would like to thank the following for permission to use photographs:

p2 T: Photodisc, BL: FLPA, BR: Corbis; p3 PA Photos; p4 Empics; p7 SPL; p8 SPL; p13 T: SPL, B: Empics; p19 SPL; p20 Empics; p21 Empics; p22 SPL; p29 Janet Tomlin; p33 SPL; p35 Empics; p38 T: PA Photos, B: Empics; p42 L & R: SPL; p43 L: Corbis, R: Empics; p44 Mary Evans Picture Library; p44 Empics; p46 PA Photos; p52 Alamy; p54 SPL; p57 L & R: SPL; p59 SPL; p72 SPL; p76 SPL; p80 SPL; p82 SPL; p83 SPL; p85 SPL; p88 Florence Low; p89 L: Haddon Davies; R: SPL; p90 T: SPL, BL: Simpsoncrazy/Matt Groening, BM: Corbis, BR: Chris Purday; p91 Corbis, p95 From "Psychology" 2nd Edn by Lucy T Benjamin Jr, J Roy Hopkins and Jack Nation. Copyright 1990, Macmillan Publishing; p102 L: Alamy, R: SPL; p105 Corbis.

T=Top, B=Bottom, L=Left, R=Right
SPL=Science Photo Library, FLPA=Frank Lane Picture Agency

Every effort has been made to contact copyright holders of material reproduced in this book. Any omissions will be rectified in subsequent printings if notice is given to the publishers.

Picture research by Pete Morris

Unit 5

Contributors

Many people from schools, colleges, universities, industries and the professions have contributed to the Salters-Nuffield Advanced Biology project. They include the following.

Central team

Angela Hall, Nuffield Curriculum Projects Centre
Michael Reiss (Director), Institute of Education, University of London
Anne Scott, University of York Science Education Group
Catherine Rowell, University of York Science Education Group
Sarah Codrington, Nuffield Curriculum Projects Centre
Nancy Newton (Secretary), University of York Science Education Group

Advisory Committee

Professor R McNeill Alexander FRS	University of Leeds
Dr Allan Baxter	GlaxoSmithKline
Professor Sir Tom Blundell FRS (Chair)	University of Cambridge
Professor Kay Davies CBE	University of Oxford
Professor Sir John Krebs FRS	Food Standards Agency
Professor John Lawton FRS	Natural Environment Research Council
Professor Peter Lillford CBE	University of York
Dr Roger Lock	University of Birmingham
Professor Angela McFarlane	University of Bristol
Dr Alan Munro	University of Cambridge
Professor Lord Winston	Imperial College, London
Dr Roger Barker	University of Cambridge

Unit 5 authors

Topic 8

Jacquie Punter	Brighton & Hove Sixth Form College
David Greenwood	Greenhead College, Huddersfield
Nick Owens	Oundle School, Peterborough
Catherine Rowell	University of York Science Education Group

Topic 9

Gill Hickman	Ringwood School
Ginny Hales	Cambridge Regional College
Steve Hall	King Edward VI School, Southampton
Jenny Owens	Rye St Antony School, Headington
Mark Winterbottom	King Edward VI School, Bury St Edmunds
Nan Davies	

Other materials
Pat Burdett

We would also like to thank the following for their advice and assistance.

John Holman	University of York
Andrew Hunt	Nuffield Curriculum Projects Centre
Jenny Lewis	University of Leeds

Sponsors

The Salters Institute
The Nuffield Foundation
The Wellcome Trust
ICI plc
Boots plc
Pfizer Limited
Zeneca Agrochemicals
The Royal Society of Chemistry

Contents

How to use this book   **iv**

Topic 8 – Run for your life   **1**
8.1 Getting moving   4
8.2 Energy for action   12
8.3 Peak performance   24
8.4 Breaking out in a sweat   34
8.5 Overdoing it   38
8.6 Improving on nature   43
Summary   **49**

Topic 9 – Grey matter   **51**
9.1 How did Kenge see the image?   54
9.2 Reception of stimuli – how does light trigger nerve impulses?   70
9.3 Regions of the brain   75
9.4 Visual development   83
9.5 Making sense of what we see   88
9.6 Learning and memory   96
9.7 Problems with the synapses   106
9.8 Genes and the brain   110
Summary   **116**

Answers Topic 8   **118**
Answers Topic 9   **121**

How to use this book

Welcome to unit five of the new Salters-Nuffield Advanced Biology course. This is the second of two student textbooks for the A2 course.

This book covers two topics: Topic 8 "Run for your life" and Topic 9 "Grey matter". In each topic we begin with a context and then draw out the underlying biological concepts. Within each topic you will develop your knowledge and understanding of these biological concepts.

Each topic contains a number of features which we hope will help your learning.

Main text

The **text** presents the contexts of the unit and explains the relevant biology. Key terms are printed in **bold**. We advise that you check you understand the meanings of these words.

The main text also contains occasional boxes. As their name suggests, **'Key biological principle' boxes** highlight some fundamental biological principles. We also have **'Nice to know' boxes**. These contain material that you won't be examined on but we hope you will find interesting.

At the end of the main text for each topic you will find a **summary**. The summary contains a bullet-pointed list of what you need to have learnt from the topic for the unit exam.

Activities

Throughout the unit you will find references to activities. Activities include practicals, interactive on-line exercises, role-plays, issues for debate, etc. Details of all the activities can be accessed on-line and most can also be used in a paper form. Your teacher or lecturer will guide you as to which activities to do and when.

Each topic finishes with an end-of-topic text. This allows you to see how much of the topic you have understood and remembered. The end-of-topic test can be completed interactively using the website.

Skills support

As part of the dedicated website you can find support on related science, maths and ICT skills.

Questions

Every now and again in the text we have provided questions. These are intended to get you to think about the material. Answers are provided at the end of the book. We'll leave it up to you to decide whether or not the best strategy is to think about the questions before you look at the answers. . . .

Any comments?

Finally, this is a pilot. We are therefore keen to hear from you with suggestions about how the course could be improved. You can e-mail us at contact@snab.org.uk

or write to us at

The Salters-Nuffield Advanced Biology Project, Science Education Group, University of York, Heslington, York YO10 5DD

Best of luck!

Run for your life

Topic 8

Run for your life

Why a topic called *Run for your life*?

▲ **Figure 8.1** The fastest land mammal, the cheetah, at full stretch.

Cheetahs can run at speeds in excess of 100 km hr^{-1} but after just a few hundred metres they must stop and rest or risk collapsing. Wildebeest can run, though not as quickly, for many kilometres. Both animals are literally running for their lives. Whether chasing prey or seeking out new grazing pasture, their survival depends on their ability to run.

Before humans began farming just 10 000–12 000 years ago we all lived as hunter-gatherers, working from a temporary home base, hunting wild animals and gathering plants from our natural surroundings. There would be long periods of moderate exercise while plucking shoots and berries, with occasional vigorous activity such as chasing after prey with spears.

Although few humans have to chase down prey or cover huge distances on foot any more, many of us still run. Exercise that involves running, or at least jogging, helps maintain our health, and for the professional

▲ **Figure 8.2** Wildebeest travel huge distances during their annual migrations to find fresh pasture.

▲ **Figure 8.3** Our abilities have evolved over thousands of years as we adapted to our surroundings.

sportsperson it also provides a living. We marvel at those who, like Paula Radcliffe in Figure 8.4, can complete a 43 km (26 mile) marathon in around two and a quarter hours and are truly amazed by the ultra-marathon runners who cover 100 km, keeping going for as long as ten hours. But how is it that the marathon runner who can complete 43 km at about a 20 km an hour pace would fail to run 100 metres in the 10 seconds that it takes the top class sprinters?

The cheetah, wildebeest, we humans and all other mammals share the same basic structures that allow mammals to move about. Bones, joints and muscles are very similar, both macroscopically and microscopically, and all mammals, indeed most animals, use the same biochemical pathways to make energy available for movement. But if we are all so similar, how is that the wildebeest and caribou can keep going for hours on end but the cheetah must rest after just a few minutes? Why do some people excel in sprint events whilst others make outstanding distance runners? Why is it so rare for one person to achieve the highest level in both?

▲ **Figure 8.4** The London Marathon in 2002 saw new world records for marathon running. Paula Radcliffe won the women's race in 2:15:25.

Exercise poses some real challenges for the body. Many changes occur without our conscious thought, such as the finely controlled adjustments required to ensure that oxygen and fuel are supplied in sufficient quantities to the muscles. Fit sportspeople may not be aware of this unless working very hard, when the heart starts pounding and breathing becomes laboured. Overheating may be something they are much more aware of as they cover the kilometres. How does the body ensure that muscles are well supplied with oxygen and fuel, and prevent body temperature from rising too high?

We are always being encouraged to take plenty of exercise. This is good advice with our increasingly sedentary lifestyles which are leading to an ever higher incidence of obesity, cardiovascular disease and related diseases. But can you overdo it? What are the consequences of overtraining or trying to improve on nature?

 Overview of the biological principles covered in this topic

At the start of the topic you recall ideas from KS4 about joints and movement. This is then extended to provide a detailed understanding of the mechanism of contraction in skeletal muscle. You revisit ideas covered in Topic 6 about energy transfer in biological systems, ATP, redox reactions and electron transfer chains in the context of respiration. The different energy systems are considered in detail. Building on your knowledge of the cardiovascular and ventilation systems from Topics 1 and 2 you investigate the control of heart rate, lung volumes and breathing rate.

In this topic homeostasis and negative feedback as illustrated by thermoregulation, discussed in Topic 7, are re-examined in a sporting context. You return to the principles of the immune response introduced in Topic 7 in the context of immune suppression as a consequence of overtraining. You investigate the use of medical technology to enable participation of those with injuries or disabilities. You will discuss the moral and ethical issues surrounding the use of performance-enhancing substances in sport.

Topic 8

8.1 Getting moving

Before the starter's gun fires, the athlete (Figure 8.5) is poised for action. He is crouched in the sprint start position on the blocks, legs and ankles flexed. As the gun fires he pushes against the blocks to get maximum thrust and is up and running. Like the cheetah, with every step, muscles contract and relax to bend (**flex**) and straighten (**extend**) ankles, knees and hip joints, exerting a force against on the ground to push his body forward.

Muscles bring about movement at a **joint**; most movements are produced by the coordinated action of several muscles. Muscles shorten, pulling on the bone and so moving the joint. Muscles can only pull. They cannot push, so at least two muscles are needed to move a bone to and fro. A pair of muscles that work in this way are described as **antagonistic**. For example, when you flex your knee by contracting the hamstring muscles at the back of the thigh, the quads at the front of the thigh relax and so are stretched. To extend the knee the quadriceps shorten while the hamstrings relax.

A muscle that contracts to cause extension of a joint is called an **extensor**; the corresponding **flexor** muscle contracts to reverse the movement.

▲ **Figure 8.5** An athlete prepared for action.

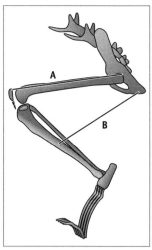

▲ **Figure 8.6** Extension and flexion of a cat knee joint.

Q8.1 In Figure 8.6 which muscle, A or B, brings about flexion of the cat knee joint?

The hip, knee and ankle joints are known as **synovial joints**; the bones that articulate in the joint are separated by a cavity filled with **synovial fluid** which enables them to move freely. All synovial joints have the same basic structure as shown in Figure 8.7. The bones are held in position by **ligaments** that control and restrict the amount of movement in the joint. **Tendons** attach muscles to the bones, enabling the muscles to power joint movement.

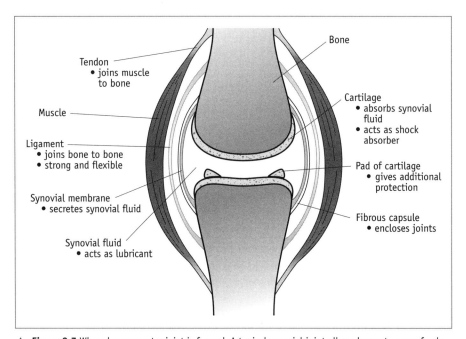

▲ **Figure 8.7** Where bones meet a joint is formed. A typical synovial joint allows bones to move freely.

Q8.2 a) What properties of ligaments make them effective at holding the bones in place at a joint?
b) What reduces wear and tear through friction in a mobile synovial joint?

> ● **Nice to know:** Other types of joints
>
> Not all joints have the same structure. The plates of the skull are fixed together by fibrous tissue so very little movement occurs at these **fixed joints**. As a result the brain is protected inside a rigid bone casket. In the skull of a newborn baby the joints are not yet fixed; this allows the plates to overlap and the skull to deform (reversibly!) during birth. There are spaces between the skull bones, which fill in as the plates grow and fuse together. A mother's pelvic bones are joined together by cartilage, so only slight movement is possible during childbirth; this is an example of a **cartilaginous joint**. There are also a wide variety of synovial joints allowing different degrees of extension, flexion and rotation, as shown in Figure 8.8.
>
>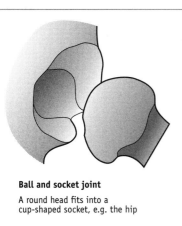
>
> **Ball and socket joint**
> A round head fits into a cup-shaped socket, e.g. the hip
>
>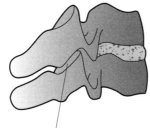
>
> **Gliding joint**
> Two flat surfaces slide over one another, e.g. the articular process between neighbouring vertebrae
>
>
>
> **Hinge joint**
> A convex surface fits into a concave surface, e.g. the elbow
>
>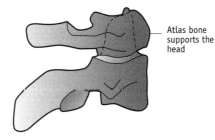
>
> Atlas bone supports the head
>
> **Pivot joint**
> Part of one bone fits into a ring-shaped structure and allows rotation, e.g. the joint at the top of the spine
>
> There are also saddle joints and condyloid joints in which more complex concave and convex surfaces articulate.

▲ **Figure 8.8** Different types of synovial joint.

Topic 8

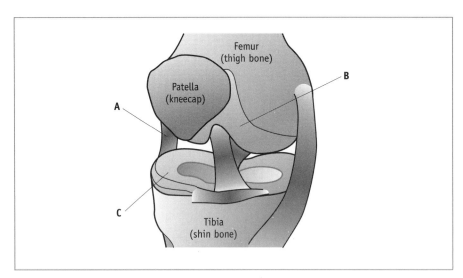

▲ **Figure 8.9** The human knee joint.

Q8.3 Compare the simplified synovial joint in Figure 8.7 with this diagram of the human knee joint (Figure 8.9).
a) Identify the key features of the knee joint labelled A to C.
b) Which features of a synovial joint are not shown in this diagram?

 Activity

You can further your understanding of the knee joint in interactive **Activity 8.1. A208ACT01**

 Nice to know: What is it about cheetahs' skeletons and claws that helps make them so fast?

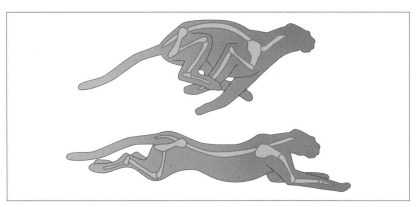

▲ **Figure 8.10** The cheetah flexes its spine.

The cheetah's slender body, long legs and flexible spine allow it run at great speed. The cheetah flexes its spine and rotates its shoulder blades forward and back which gives it a longer stride (Figure 8.10). Unlike other cats, the cheetah does not retract its claws into protective sheaths; the exposed claws, combined with hard ridges on their feet pads, function rather like an athlete's spikes. The disadvantage of having permanently exposed claws is that they become blunt, making them less good as running spikes and for catching prey. The latter is not too much of a problem for the cheetah which uses an enlarged claw on the inner sides of the front legs to bring down moving prey.

How do muscles work?

Look at Figure 8.11 which shows some human muscle cells (**fibres**). Apart from the considerable length of the cells, what do you notice about the muscle fibres that is unusual? You should spot that each cell has several nuclei. This occurs because a single nucleus could not effectively control the metabolism of such a long cell. During prenatal development, several cells fuse together forming an elongated muscle fibre. The muscle cells are also striped; as we will see, this is an important feature related to their ability to contract.

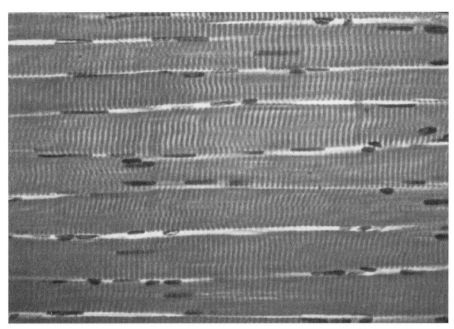

▲ **Figure 8.11** Human muscle fibres. Each fibre is a single cell that runs the whole length of the photograph.

Q8.4 Why would a single nucleus be unable to control the metabolism of a long thin cell like a muscle fibre?

What is going on inside a muscle fibre?

The key to how muscles function is in their internal structure (Figures 8.12 and 8.13). Contractions are brought about by co-ordinated sliding of protein **filaments** within the muscle cells.

The sarcomere is made up two types of protein molecules: thin filaments made up mainly of **actin** and thicker ones made of **myosin**.

The arrangement of the filaments is shown in Figures 8.13 and 8.14. These proteins overlap and give the muscle fibre its characteristic striped or **striated** appearance under the microscope (Figure 8.11). Where actin filaments occur on their own, there is a light band on the sarcomere. Where both actin and myosin filaments occur there is a dark band. Where only myosin filaments occurs there is an intermediate grey band.

> **Weblink**
>
> Compare these muscle cells with the animal cell in the interactive cell, Activity 3.04

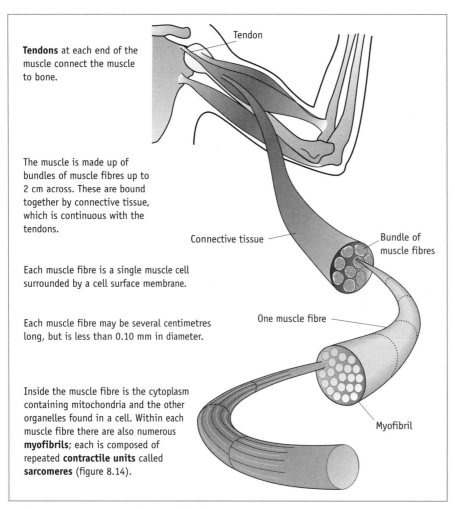

▲ **Figure 8.12** Arrangement of muscle fibres within muscles.

Text within figure 8.12:

Tendons at each end of the muscle connect the muscle to bone.

The muscle is made up of bundles of muscle fibres up to 2 cm across. These are bound together by connective tissue, which is continuous with the tendons.

Each muscle fibre is a single muscle cell surrounded by a cell surface membrane.

Each muscle fibre may be several centimetres long, but is less than 0.10 mm in diameter.

Inside the muscle fibre is the cytoplasm containing mitochondria and the other organelles found in a cell. Within each muscle fibre there are also numerous **myofibrils**; each is composed of repeated **contractile units** called **sarcomeres** (figure 8.14).

▲ **Figure 8.13** Electron micrograph showing the banding pattern of a sarcomere.

When the muscle contracts, the actin moves between the myosin as shown in Figure 8.14(c), shortening the length of the sarcomere and hence the length of the muscle.

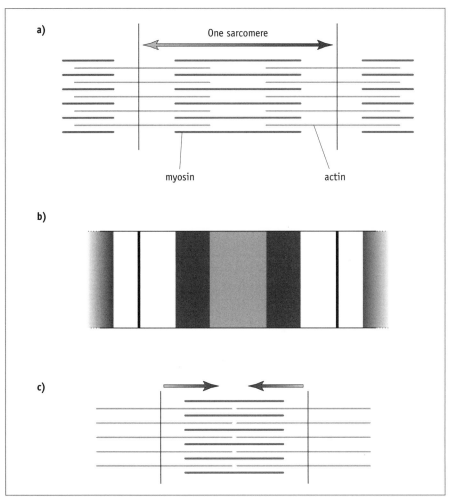

▲ **Figure 8.14** a) The arrangement of actin and myosin filaments within a single muscle sarcomere when relaxed; b) The banding patterns created on a relaxed muscle myofibril; c) The arrangement when the muscle contracts.

Q8.5 Look at the banding pattern on the contracted muscle fibre shown below. Explain what has happened to the central grey band visible on uncontracted muscle.

▲ **Figure 8.15** The banding pattern of a sarcomere when contracted.

> **Activity**
>
> **Activity 8.2** will let you investigate internal muscle structure, and see how muscle shortens during contraction.
> **A208ACT02**

Topic 8

How actin moves – the sliding filament theory

Myosin molecules are shaped rather like golf clubs; the club shafts lie together as a bundle, with the heads protruding along their length. Movement of these myosin heads brings about the movement of actin, and hence shortening of the muscle sarcomere – see Figure 8.16.

When a nerve impulse arrives at a **neuro-muscular junction**, calcium ions (Ca^{2+}) are released into the **sarcoplasm** from the **sarcoplasmic reticulum**. This is a specialised type of endoplasmic reticulum, a system of membrane-bound sacs around the myofibrils, Figure 8.16a.

Actin molecules are associated with two other protein molecules called **troponin** and **tropomyosin**. The Ca^{2+} ions attach to the troponin molecules, causing them to move. As a result, the tropomyosin on the actin filament shifts its position, exposing myosin binding sites on the actin filaments.

The myosin heads bind with myosin binding sites on the actin filament, forming **cross bridges**. When the myosin head binds to the actin, ADP and Pi on the myosin head are released.

The myosin changes shape, causing the myosin head to nod forward. This movement results in the relative movement of the filaments.

An ATP molecule binds to the myosin head. An ATPase on the myosin head hydrolyses the ATP, forming ADP and Pi. This causes a change in shape of the myosin head, allowing the head to detach from the actin and return to its upright resting position, and enabling the cycle to start again.

In a contraction, when the muscle shortens, the change in orientation of the myosin head causes the attached actin filament to slide over the myosin. This is called the **sliding filament theory**.

At any one time during muscle contraction about half the myosin heads are attached to the actin, while the others are recovering. The number of heads attached falls as the rate of shortening increases. In the absence of ATP the cross bridges remain attached and the muscle is contracted. This is what happens in rigor mortis, described in Topic 7, page 79.

When a muscle relaxes it is no longer being stimulated by a nerve impulse. Calcium ions are actively pumped out of the muscle sarcoplasm, using ATP. The troponin and tropomyosin move back, once again blocking the actin binding site.

 Activity

Building a model in **Activity 8.3** will let you check your understanding of the sliding filament theory. **A208ACT03**

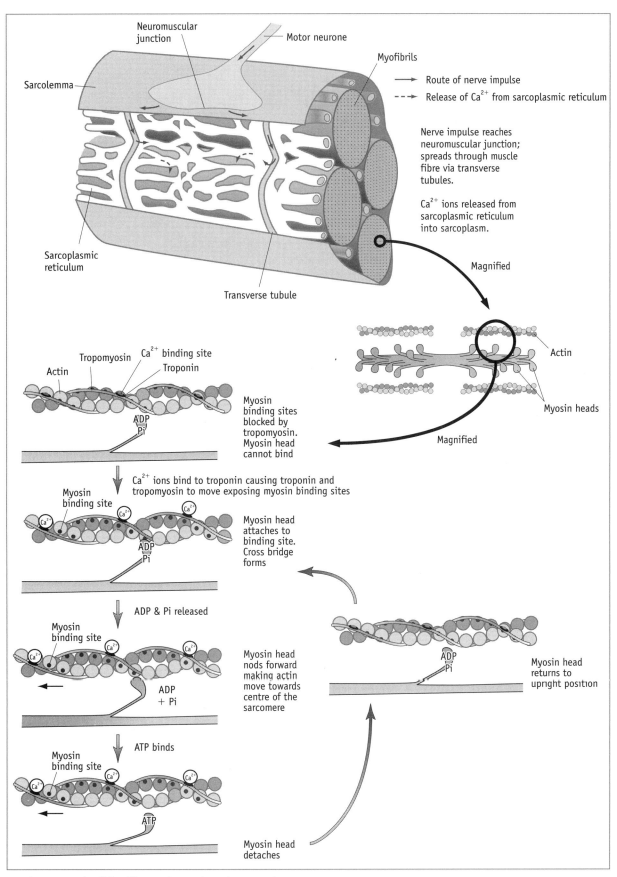

▲ **Figure 8.16** The sliding filament theory of muscle contraction.

Topic 8

> **Extension**
>
> Find out what the method of wrinkle control (Botox), and a lethal form of food poisoning (botulism) have in common in **Extension 8.1**. **A208EXT01**

Muscles found in such places as the gut wall, blood vessels and iris of the eye are known as **smooth muscle**: their fibres do not appear striped. They have a similar mechanism of contraction using myosin and actin protein filaments. However, smooth muscle fibres contract and fatigue very slowly if at all.

The heart walls are made of specialised **cardiac muscle** fibres; these are striped and interconnected to ensure that a coordinated wave of contraction occurs. Cardiac muscle fibres do not fatigue. Both smooth and cardiac muscle are not under conscious control – unless you became extremely good at certain forms of yoga.

8.2 Energy for action

Just staying alive, even if you're not doing anything active, uses a considerable amount of energy. This minimum energy requirement is called the **basal metabolic rate** (**BMR**). (See Topic 1 in student book 1 page 37.) BMR is a measure of the minimum energy requirement of the body at rest to fuel basic metabolic processes.

BMR is measured by recording oxygen consumption under strict conditions; no food is consumed for 12 hours before measurement with the body totally at rest in a thermostatically-controlled room. BMR is roughly proportional to the body's surface area. It also varies between individuals depending on their age and gender. Percentage body fat seems to be important in accounting for these differences.

Physical activity increases the body's **total daily energy expenditure**. Energy is needed for muscle contraction to move the body around (see Activity 1.19). An elite marathon runner expends energy at a rate of about 1.75 kJ per second for the duration of the run.

Food is the source of energy for all animal activity. The main energy sources for most people are carbohydrates and fats which have either just been absorbed from the gut or have been stored around the body. A series of enzyme-controlled reactions are linked to ATP synthesis. As we saw in Topic 6, cells use the molecule ATP as an energy carrier molecule. This is the cells' energy currency, coupling energy-yielding reactions and energy-requiring reactions like the muscle contraction described earlier in this topic.

When phosphate groups are removed from ATP in hydrolysis, forming adenosine diphosphate, some of the energy transferred will raise the temperature of the surrounding tissues; some is available to drive other metabolic reactions such as muscle contraction, protein synthesis or active transport. The hydrolysis of ATP is *coupled* to these other reactions:

$$ATP \rightarrow ADP + Pi + energy$$

Q8.6 Women generally have a higher percentage of body fat than men. Why might this account for the fact that women generally have low BMR?

▲ **Figure 8.17** ATP is a universal molecule used by living organisms. Here fireflies release energy from ATP to create luminescence in their attempts to attract mates.

Recharging the batteries

Cells store only a tiny amount of ATP, enough in humans to allow a couple of seconds of explosive, all-out exercise. The recharging of the ATP store therefore has to be very rapid.

At the start of any type of exercise the immediate regeneration of the ATP is achieved using **creatine phosphate** (sometimes also called phosphocreatine, PC). This is a substance stored in muscles which can be hydrolysed to release energy – this energy can be used to regenerate ATP from ADP and phosphate, the phosphate being provided by the creatine phosphate itself. Creatine phosphate breakdown begins as soon as exercise starts (triggered by the formation of ADP). It does not require oxygen, and provides energy for about 6–10 seconds. This is known as the **ATP/PC** system and is relied upon for regeneration of ATP during bursts of intense activity, for example when throwing or sprinting (Figure 8.18). This can be represented using the linked reactions:

Creatine phosphate → Creatine + P_i

ADP + P_i → ATP

These can be summarised as:

Creatine phosphate + ADP → Creatine + ATP

Later, creatine phosphate stores can be regenerated from ATP when the body is at rest.

If exercise is low intensity, oxygen supply to cells is sufficient to enable ATP to be regenerated through aerobic respiration of fuels. Fats and carbohydrates, like glucose, are oxidised to carbon dioxide and water; you are probably familiar with the summary equation for aerobic respiration:

$C_6H_{12}O_6$ + $6H_2O$ → $6CO_2$ + $6H_2O$ + energy released

▲ **Figure 8.18** These sprinters rely almost entirely on the ATP/PC system to supply the energy burst for their 10 second race.

Energy released when fuel is oxidised is used to generate ATP. This process is the most efficient method of regenerating ATP, i.e. of obtaining the maximum yield of ATP molecules per glucose molecule.

Carbohydrate oxidation – glycolysis first

A series of enzyme-controlled reactions have evolved for the oxidation of carbohydrates; a sequence of small reactions ensures that there is a controlled release of energy which avoids overheating the cell. The initial stages of carbohydrate breakdown, known as **glycolysis**, occur in the cytoplasm of cells, including the sarcoplasm of muscle cells.

Stores of glycogen (a polymer of glucose, see Topic 1) in muscle or liver cells are converted to glucose. Glucose is a good fuel – it can potentially yield 2880 kJ mol^{-1} – but it is quite stable and unreactive. The first reactions of glycolysis need an *input* of energy from ATP to get things started. Two phosphate groups are added to the glucose from two ATP molecules, and this increases the reactivity of the glucose. It can now be split into two molecules of 3-carbon (3-C) compounds. It is rather like lighting a candle; a match provides an initial input of energy before energy can be released from the fuel (candle wax).

Each 3-C sugar is oxidised, producing the 3-carbon compound, **pyruvate**; two hydrogen atoms are also removed during the reaction. The fate of these hydrogens and their role in ATP synthesis is covered on page 18. Glucose is at a higher energy level than the pyruvate so, on conversion, some energy becomes available for the direct creation of ATP. Phosphate from the intermediate compounds is transferred to ADP, creating ATP. This is called **substrate-level phosphorylation**, because energy for the formation of ATP comes from the substrates; in this case the intermediate compounds are the substrates (Figure 8.19).

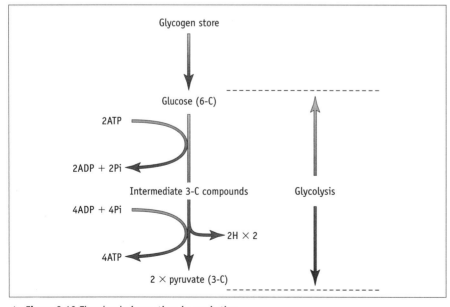

▲ **Figure 8.19** The glycolysis reactions in respiration.

In summary, glycolysis reactions yield a net gain of 2ATPs, 2 hydrogen atoms, and two molecules of 3-carbon pyruvate as shown in Figure 8.19.

Q8.7 Are the reactions in glycolysis aerobic or anaerobic? Give a reason for your answer.

What happens to the pyruvate depends on the availability of oxygen.

The fate of pyruvate if oxygen is available

If oxygen is available, the 3-C pyruvate created at the end of glycolysis passes into the mitochondria. There it is completely oxidised forming three carbon dioxide molecules and water (see Figure 8.20 and 8.22).

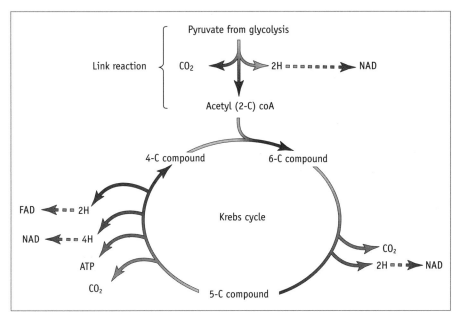

▲ **Figure 8.20** The reactions involved in the breakdown of glucose in aerobic respiration. Each glucose provides two pyruvates so the cycle turns twice per glucose.

The first step is that pyruvate is decarboxylated (carbon dioxide is released as a waste product), dehydrogenated (two hydrogens are removed) and the resulting two-carbon molecule combined with coenzyme A to form **acetyl coenzyme A** (**acetyl coA** for short). As we shall see, the two hydrogen atoms released are involved in ATP formation. The coenzyme A carries the 2-C acetyl groups to the Krebs cycle. The reaction of pyruvate with acetyl coA is known as the link reaction.

Krebs cycle

The 2-C compound carried by each acetyl coA combines with a 4-C compound to create one with six carbons. In a circular pathway of reactions, the original 4-carbon compound is recreated. In these reactions, two steps involve decarboxylation with the formation of carbon dioxide. Four steps involve dehydrogenation, the removal of pairs of hydrogen atoms. In addition, one of the steps in the cycle also involves substrate-level phosphorylation with direct synthesis of a single ATP (see glycolysis substrate-level phosphorylation on page 14). This circular pathway of reactions (Figure 8.20) is known at the **Krebs cycle**, named after Sir Hans Krebs who worked out the cycle of reactions. The Krebs cycle takes place in the mitochondrial matrix (Figure 8.23) where the enzymes that catalyse the reactions are located.

Key biological principle: Understanding the chemistry of respiration

The chemical reactions inside cells are controlled by enzymes. There are four important types of reactions in the Krebs cycle:

- Phosphorylation reactions which add phosphate
  e.g. ADP + Pi → ATP
- Decarboxylation reactions which break off carbon dioxide
  e.g. pyruvate → acetyl coA + CO_2
- Dehydrogenation reactions which remove hydrogen
  e.g. pyruvate → acetyl coA + 2H
- Redox reactions where substrates are oxidised and reduced
  e.g. oxidised NAD + electrons → reduced NAD

When a molecule is oxidised, electrons, e^-, are lost from the molecule and another molecule which accepts these electrons is reduced. See Topic 6, page 27.

Activity

In **Activity 8.4** you can produce a summary diagram of glycolysis and the Krebs cycle.
A208ACT04

In summary, each 2-carbon molecule entering the Krebs cycle results in the production of two carbon dioxide molecules, one molecule of ATP by substrate-level phosphorylation, and eight pairs of hydrogen atoms. These hydrogen atoms are subsequently involved in ATP production via the **electron transport chain** described in the next section.

Extension

Extension 8.2 explores the stages in the discovery of the Krebs cycle to illustrate the reactions involved. **A208EXT02**

Fatty acid oxidation

Fats can also be respired to release energy and are a richer energy store than carbohydrates – glycogen releases 17 kJ g^{-1} whereas triglycerides release 37 kJ g^{-1}. In fatty acid oxidation, the glycerol and fatty acids which make up triglycerides are separated. The fatty acids are broken down in a series of reactions, each generating the same two-carbon compound, and these can be fed into the Krebs cycle for oxidation (Figure 8.21).

Because fatty acids can only be respired through the Krebs cycle, fats can only be a fuel for aerobic respiration, and cannot be used when oxygen is not available. Glucose can be respired aerobically or anaerobically.

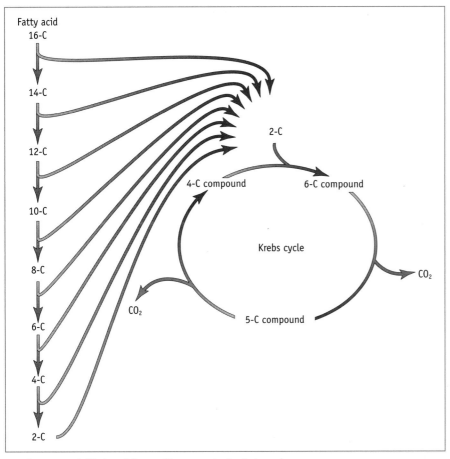

▲ **Figure 8.21** Oxidation of fatty acids occurs via the Krebs cycle.

> ● **Nice to know:** Using proteins
>
> Some modern athletes believe, like their ancient Olympian counterparts, that a high-protein diet is the key to successful competition. However, excess intake of any fuel, including protein, is converted into fat. Amino acids will only be built into muscle protein if there is training overload. The body may use the excess protein as fuel through respiration but to do this the amino acids must be first be deaminated, producing urea. Water is needed to excrete the extra urea. As urine output increases, the body's fluid requirements also increases.

Fate of the hydrogens – the electron transport chain

Hydrogen acceptors (Figure 8.23) take up the hydrogen atoms released during glycolysis, the link reaction and the Krebs cycle. For most hydrogens produced, the coenzyme **NAD (nicotinamide adenine dinucleotide)** is the **hydrogen acceptor**, although those released at one stage in the Krebs cycle are accepted by the enzyme **FAD (flavine adenine dinucleotide)** rather than NAD. When a coenzyme accepts the hydrogen with its electron it is reduced, becoming reduced NAD or reduced FAD. The reduced coenzyme 'shuttles' the electrons to the electron transport chain on the mitochondrial inner membrane. As illustrated in Figure 8.22, each hydrogen atom's electron and proton separate, with the electron passing along a chain of electron carriers in the inner mitochondrial membrane. This is known as the **electron transport system** or **electron transport chain**. As the electrons pass from one carrier to the next in a series of redox reactions, energy is released and used to generate ATP.

> **Activity**
>
> Using the context of mitochondrial diseases, **Activity 8.5** has an interactive animation to look at the electron transport chain and chemiosmosis.
> **A208ACT05**

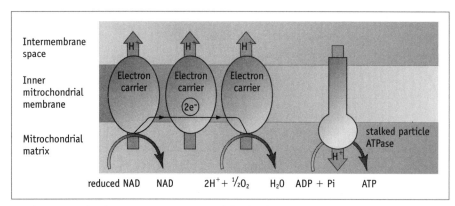

▲ **Figure 8.22** The electron transport system and the chemiosmotic theory of ATP synthesis.

Synthesis of ATP by chemiosmosis

How does the electron transport system lead to ATP synthesis? This is explained in the **chemiosmotic theory** summarised in Figure 8.22. Using energy released as electrons pass along the electron transport chain, protons (hydrogen ions) originating from the hydrogen atoms released in glycolysis or the Krebs cycle are pumped from the matrix across the inner mitochondrial membrane into the intermembrane space. This creates a steep **electrochemical gradient** across the inner membrane. There is a large difference in concentration of H^+ across the membrane, and a large electrical difference; the intermembrane space is more positive than the matrix.

The protons diffuse down this electrochemical gradient through hollow protein channels in stalked particles on the membrane. As the hydrogen passes through the channel, ATP synthesis is catalysed by ATPase located in each stalked particle. The hydrogen causes a conformational change (change in shape) in the enzyme's active site so the ADP can bind. Within the matrix, the H^+ and electrons recombine to form hydrogen atoms. These combine with oxygen to form water. The oxygen, acting as the final carrier in the electron transport system, is thus reduced. This method of synthesising ATP is known as **oxidative phosphorylation**.

Q8.8 Why is the synthesis of ATP via the electron transport system termed 'oxidative phosphorylation'?

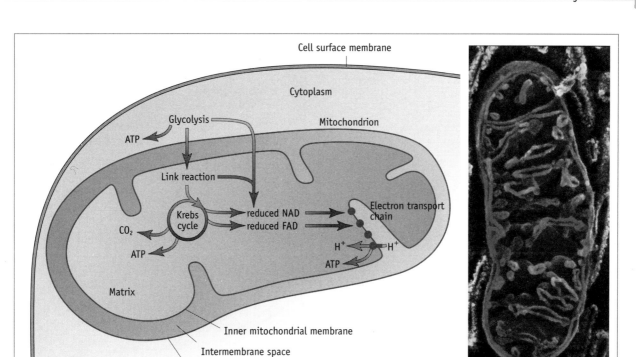

▲ **Figure 8.23** A summary of where the respiration reactions occur.

Each reduced NAD which is reoxidised makes enough energy available to produce 3ATP. Each reduced FAD transfers sufficient energy to produce 2ATP. The total number of ATP produced in the different stages of the cycle can be calculated using Figure 8.20. Try Question 8.9 to check whether you can work it out.

Q8.9 Look at Figures 8.19 and 8.20 and work out the total number of ATPs produced by a) substrate-level phosphorylation and b) oxidative phosphorylation when one glucose molecule passes through the respiration reactions.

Combining the values determined in Question 8.9, you should find that a total of 38ATP are produced from the complete aerobic respiration of each glucose molecule. Complete oxidation (e.g. by combustion) of one mole of glucose molecules releases 2880 kJ. 38 moles of ATP molecules can only release 1163 kJ of the potential chemical energy stored in the glucose, i.e. 40% of the total. The remaining energy raises the temperature of the surrounding cells; although not useful for driving cell metabolic reactions, it does maintain core body temperature.

Rates of respiration can be determined by measuring the uptake of oxygen using a respirometer (Figure 8.24).

Topic 8

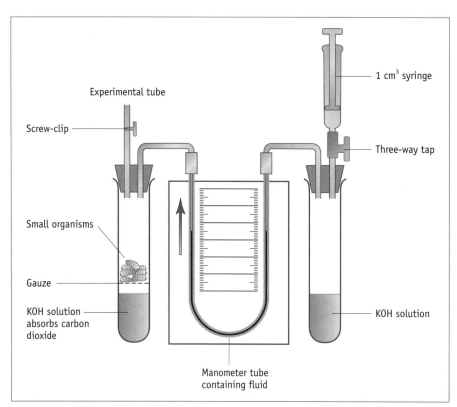

▲ **Figure 8.24** As the living organisms in the experimental tube take up oxygen the fluid in the manometer will move in the direction of the arrow. The second tube takes account of any changes in volumes due to variations in atmospheric pressure or temperature.

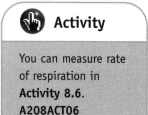

Activity

You can measure rate of respiration in **Activity 8.6.** A208ACT06

Q8.10 What is the advantage of having a system of enzyme-controlled reactions to transfer energy from food fuels?

At the start of any exercise the lungs and circulation are not delivering oxygen to muscle cells as fast as it is needed. As a result ATP will largely be synthesised without using oxygen.

Respiration without oxygen

At the start of any exercise and during intense exercise, for example a 400 m race, oxygen demands in the cells exceed supply (Figure 8.25). Without oxygen to accept the hydrogen ions and electrons, the electron transport chain ceases and the reduced NAD created during glycolysis, the link reaction and the Krebs cycle is not oxidised. Without a supply of NAD the respiration reactions cannot continue. However, it is possible to oxidise the reduced NAD created during glycolysis, using pyruvate produced at the end of glycolysis. The pyruvate is reduced to **lactate** and the oxidised form of NAD is regenerated. In this way anaerobic respiration allows the athlete to continue by partially breaking down glucose to make a small amount of ATP (Figure 8.26). The net yield is just two ATP molecules per glucose molecule – in other words, only 61 kJ energy made available in the form of ATP. This is a highly inefficient process at only 2% efficiency.

▲ **Figure 8.25** The women's world record for running 400 m is 47.6 seconds. The runner cannot rely on the ATP/PC system for energy throughout the race – it is too long and will exhaust the supplies of PC. The intensity of the activity is such that the cell does not get enough oxygen for aerobic respiration. How do such athletes fuel their performance?

The end product of anaerobic respiration is lactate, which builds up in the muscles and must be disposed of later. Animal tissues can tolerate quite high levels of lactate, but lactate forms an acid in solution. This means that as lactate accumulates, the pH of the cell falls, inhibiting the enzymes which catalyse the glycolysis reactions. The glycolysis reactions and the physical activity that depends on them cannot continue.

> **Activity**
>
> Complete the worksheet on anaerobic respiration in **Activity 8.7**.
> A208ACT07

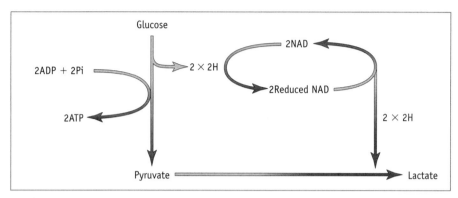

▲ **Figure 8.26** Anaerobic repiration.

Enzymes function most efficiently over a narrow pH range. Many of the amino acids that make up an enzyme have positively or negatively positively charged groups. As hydrogen ions from the dissociated acid accumulate in the cytoplasm, they will neutralise the positively charged groups in the active site of the enzyme. The attraction between charged groups on the substrate and in the active site will be affected (Figure 8.27). The substrate may no longer bind to the enzyme's active site.

▲ **Figure 8.27** As lactate builds up within cells, the pH eventually falls, inhibiting the glycolysis reactions. The athlete (or anyone sprinting for the bus!) has to slow down so that oxygen supply can meet demand. Lactate build up may lead to muscular cramp as shown in the photo above.

After a period of anaerobic respiration, most of the lactate is converted back into pyruvate and is oxidised directly to carbon dioxide and water via the Krebs cycle, thus releasing energy to synthesise ATP. As a result, oxygen uptake is greater than normal in the recovery period after exercise (see Figure 8.28). This excess oxygen requirement is called the **oxygen debt**, or post-exercise oxygen consumption. It is needed to fuel the oxidation of lactate. Some lactate may also be converted into glycogen and stored in the muscle or liver.

Topic 8

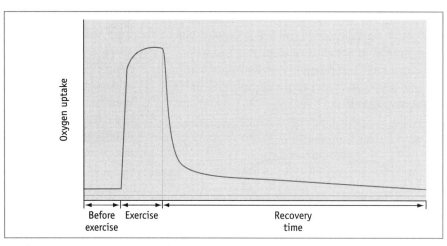

▲ **Figure 8.28** Recovery oxygen uptake.

Q8.11 Why are athletes told not to simply stop or lie down after a period of strenuous exercise, but rather to go for active recovery through gentle exercise?

> **Nice to know:** Alcoholic fermentation
>
> Yeast cells adopt a different tactic for dealing with anaerobic conditions. They reduce pyruvate to ethanol and CO_2, using the hydrogen from reduced NAD, thus recreating oxidised NAD (Figure 8.29). This anaerobic respiration, also known as **fermentation**, is much exploited in the brewing industry. Yeast are **facultative anaerobes**. **Aerobes** are organisms that respire using oxygen; **anaerobes** respire without using oxygen. Facultative anaerobes can use either aerobic or anaerobic respiration depending on the conditions.
>
>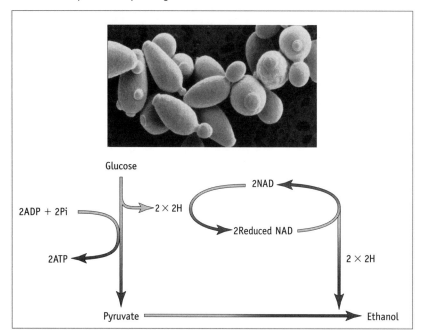
>
> ▲ **Figure 8.29** Yeast can respire both aerobically and anaerobically. In the absence of oxygen they use alcohol fermentation.

At the start of any exercise, aerobic respiration cannot meet the demands for energy because the supply of oxygen to the muscles is insufficient (Figure 8.30). The lungs and circulation are not delivering oxygen quickly enough, and ATP will be regenerated without using oxygen. First the ATP/PC system and then the anaerobic respiration system allow ATP regeneration. In endurance-type exercise, soon after the start oxygen supply to the muscle cells is sufficient for aerobic respiration to regenerate ATP as quickly as it is broken down. The relative contributions of the three different systems to the regeneration of ATP during exercise are shown in Figure 8.31.

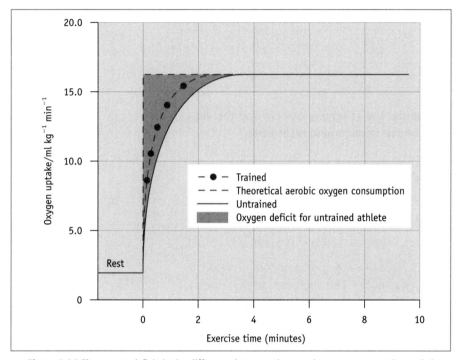

▲ **Figure 8.30** The oxygen deficit is the difference between the actual oxygen consumption and the theoretical oxygen consumption had the exercise been completed entirely aerobically.

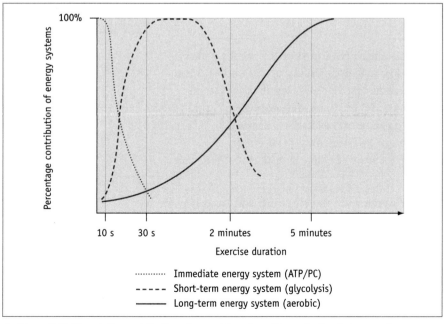

▲ **Figure 8.31** The relative contribution of anaerobic and aerobic respiration during exercise.

Topic 8

Q8.12 Will the trained or untrained athlete create a larger oxygen deficit during the exercise period shown on the graph in Figure 8.30? Give a reason for your answer.

Q8.13 Which energy system will a) a cheetah use in its sprint to catch a prey, b) a wildebeest use during the majority of its migration?

Q8.14 The table below shows the percentage contributions of the three energy systems during various sports. Which represents a) volleyball, b) hockey, and c) long distance running?

	ATP/PC	Anaerobic glycolysis to lactate	Aerobic respiration
A	60	20	20
B	90	10	0
C	10	20	70

Q8.15 A cheetah is a carnivore, so its diet is largely protein and fat. How might it gain carbohydrate stores for use in anaerobic respiration?

8.3 Peak performance

The ability to undertake prolonged periods of strenuous but submaximal activity (e.g. running but not flat out sprinting) is dependent on **aerobic capacity**, that is the ability to take in, transport and use oxygen.

At rest we consume about 0.2 to 0.3 litres of oxygen per minute. This is known as $\dot{V}O_2$. ($\dot{V}$ means volume per minute.) This increases to 3–6 litres a minute during maximal aerobic exercise, known as $\dot{V}O_2(max)$. Successful endurance athletes have a higher $\dot{V}O_2(max)$. $\dot{V}O_2(max)$ is dependent on the efficiency of uptake and delivery of oxygen by the lungs and cardiovascular system, and the efficient use of oxygen in the muscle fibres. A fit person can work for longer and at a higher intensity using aerobic respiration without accumulating lactate than can someone who does not undertake regular aerobic exercise. $\dot{V}O_2(max)$ is often expressed in units of ml min^{-1} kg^{-1} of body mass.

Most of us have run at some time and ended feeling short of breath with a pounding heart, or have seen a pet dog collapse panting at the end of a run. Without any conscious thought, the cardiovascular and ventilation systems adjust to meet the demands of the exercise, ensuring that enough oxygen and fuel reaches the muscles, and removing excess carbon dioxide and lactate. These systems are also important in redistribution of energy in temperature control. The major changes are to cardiac output, breathing rate and the depth of breathing.

Look back at Topic 1 to refresh your memory of the structure and function of the heart, and Topic 2 to remind yourself about lungs and gas exchange.

Activity

In **Activity 8.8** measure your $\dot{V}O_2(max)$.
A208ACT08

Cardiac output

The volume of blood pumped by the heart in one minute is called the **cardiac output**. This increases during exercise. At rest this is approximately 5 dm^3 (litres) per minute in both trained and untrained individuals, but it can rise to about 30 dm^3 min^{-1} in a trained athlete making maximum effort. The cardiac output depends on the volume of blood ejected from the left ventricle (the stroke volume) and the heart rate:

Cardiac output = stroke volume (SV) × heart rate (HR)

Stroke volume

The **stroke volume** is the volume of blood pumped out of the left ventricle each time the ventricle contracts, measured in cm^3. (The volume pumped from both the left and right ventricles is virtually identical. Think about it!). For most adults at rest, about 50–90 cm^3 is pumped into each of the pulmonary artery and aorta when the ventricles contract.

How much blood the heart pumps out with each contraction is determined by how much blood is filling the heart, that is the volume of blood returning to the heart from the body. The heart draws blood into the atria as it relaxes during diastole. During exercise there is greater muscle action so more blood returns to the heart in what is known as **venous return**. In diastole during exercise the heart fills with a larger volume of blood. The heart muscle is stretched to a greater extent, causing it to contract with a greater force, and so more blood is expelled, increasing stroke volume. As stroke volume increases so does cardiac output.

At rest the ventricles do not completely empty with each beat; approximately 40% of the blood volume remains in the ventricles after contraction. During exercise stronger contractions occur, ejecting more of the residual blood from the heart. In Figure 8.32 you can see the effect of exercise on stroke volume.

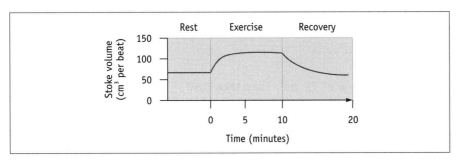

▲ **Figure 8.32** The effect of exercise on stroke volume.

Heart rate

Each of us has a slightly different resting heart rate. Measure your heart rate by counting your pulse while sitting at your desk. With each beat of your heart, a pulse of blood is ejected. This can be felt passing along the arteries; you can feel it fairly easily at the wrist (radial artery) or at your neck (carotid artery). The average heart rate for males is 70 beats per minute (bpm) while for females it is 72 bpm. The average fit person's heart rate would be around 65 bpm.

 Activity

Activity 8.9 lets you look at the effect of exercise on cardiac output. **A208ACT09**

Topic 8

Differences in your resting heart rate are caused by many factors. For example, we differ in the size of the heart, owing to differences in body size and genetic factors. A larger heart will expel more blood with each beat and so, other things being equal, does not have to beat as frequently to circulate the same volume of blood round the body. Endurance training produces a lower resting heart rate, largely due to an increase in the size of the heart, resulting from thickening of the heart muscle walls. The cyclist Miguel Indurain, five times winner of the Tour de France, had a resting pulse rate of 28 beats per minute.

Look at the graphs in Figure 8.33 to see the effect of exercise on heart rate and cardiac output.

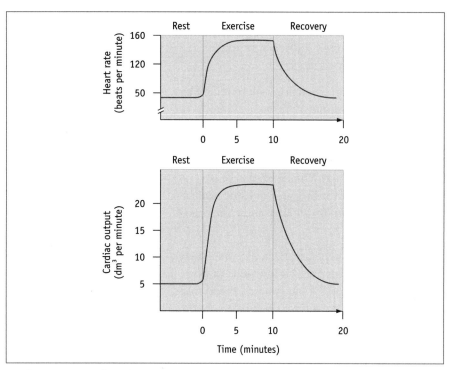

▲ **Figure 8.33** The effect of exercise on heart rate and cardiac output.

Q8.16 a) A person has a resting stroke volume of 75 cm^3. They take their pulse rate and find that it is 70 beats per minute. What is their cardiac output?
b) An endurance athlete has the same cardiac output at rest, but has a resting heart rate of 50 bpm. What is their stroke volume?
c) The cyclist Miguel Indurain's resting heart rate is 28 bpm. What will his resting stroke volume be assuming his resting cardiac output is much the same as everyone elses'?

Control of heart rate

Remind yourself about the cardiac cycle by reviewing Topic 1 pages 11 and 12.

The heart can beat even when it is removed from the body and placed in glucose and salt solution. This shows that the heart muscle is **myogenic**; it can contract without external nervous stimulation. However, the heart rate is under the control of the **cardiovascular control centre** located in the medulla of the brain. This control centre detects accumulation of carbon dioxide and lactate in the blood, reduction of oxygen, and increased temperature. Mechanical activity in muscles and joints is detected by sensory receptors in muscles, and impulses are sent to the cardiovascular control centre.

These changes result in higher heart rate. Nerves that form part of the **autonomic nervous system** (the nervous system which you have no control over) lead from the cardiovascular control centre to the heart. There are two such nerves going from the cardiovascular control centre to the heart – a sympathetic (accelerator) nerve and the vagus nerve which is a parasympathetic nerve and acts as a decelerator. See Table 8.1 for a comparison of the functions of these different nerve types. Stimulation of the sino-atrial node (SAN) by the sympathetic nerve causes an increase in the heart rate, whereas impulses from the vagus nerve slow down the rate (Figure 8.34).

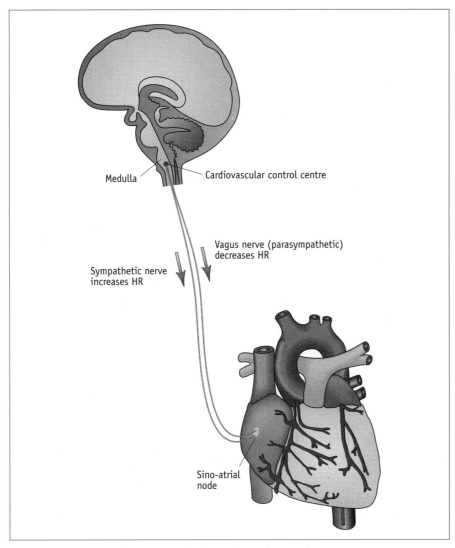

▲ **Figure 8.34** Control of the heart rate by the cardiovascular control centre.

Topic 8

▼ **Table 8.1** Sympathetic and parasympathetic nerves.

The autonomic (unconscious) part of your nervous system is made up of two sets of nerves:
• **Sympathetic** – stimulation of the sympathetic nerves prepares the body's systems for action (for the fight or flight response).
• **Parasympathetic** – stimulation of the parasympathetic nerves control the body's systems when resting and digesting.
See examples of the opposing effects of the two types of nerve below:

Organ or tissue	Effect of sympathetic stimulation	Effect of parasympathetic stimulation
Intercostal muscles	Increases breathing rate	Decreases breathing rate
Heart	Increases heart rate and stroke volume	Decreases heart rate and stroke volume
Gut	Inhibits peristalsis	Stimulates peristalsis

Fear, excitement and shock cause the release of the hormone adrenaline into the blood stream from the adrenal glands located above the kidneys. Adrenaline has a similar effect on the heart rate as do stimulations by the sympathetic nerves. It has a direct effect on the sino-atrial node, increasing the heart rate to prepare the body for any likely physical demands. Adrenaline also causes dilation of the arterioles supplying skeletal muscles, and constriction of arterioles going to the digestive system and other non-essential organs; this maximises blood flow to the active muscles. Before the start of a race, adrenaline causes an anticipatory increase in heart rate.

At the sound of the starting pistol, or the sight of prey (an animal's next meal), skeletal muscles contract, and stretch receptors in the muscles and tendons are stimulated and send impulses to the cardiovascular control centre. This in turn raises the heart rate via the sympathetic (accelerator) nerve. There is an increase in venous return which leads to a rise in the stroke volume. Together, the elevated heart rate and stoke volume result in higher cardiac output, thus transporting oxygen and fuel to muscles more quickly.

Blood pressure rises with higher cardiac output. To prevent it rising too far, pressure receptors in the aorta and in the carotid artery send nerve impulses back to the cardiovascular control centre. Inhibitory nerve impulses are then sent from here to the sino-atrial node, so an excessive rise in blood pressure is prevented by negative feedback, preventing further rise in heart rate.

Q8.17 The carotid artery at the side of the neck is sometimes used to measure heart rate (pulse rate).
a) Suggest why pressing on the carotid artery might reduce pulse rate, thereby giving a false reading.
b) Where could you take a pulse more reliably?

Q8.18 Look at Figure 8.33. Explain why the heart rate rises before the start of exercise and why this may be an advantage for the animal.

Run for your life

Breath-taking

The volume of air we breathe in and out at each breath is our **tidal volume**. At rest this is usually about 0.5 dm³. When exercise begins we increase our breathing rate and depth of breathing. The maximum volume of air we can **inhale** and **exhale** is our **vital capacity**. In most people this is 3–4 dm³, but in large or very fit people it can be 5 dm³ or more. Singers and those playing wind instruments may also have a large vital capacity. Lung volumes (tidal volume, vital capacity, etc) can be measured using a spirometer (see Figure 8.35 and Activity 8.10).

> **Activity**
>
> In **Activity 8.10** you can measure lung volumes and breathing rate.
> **A208ACT10**

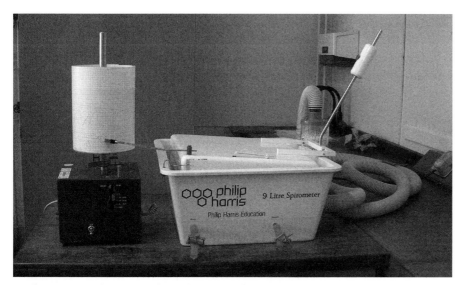

▲ **Figure 8.35** A spirometer can be used to measure lung volumes.

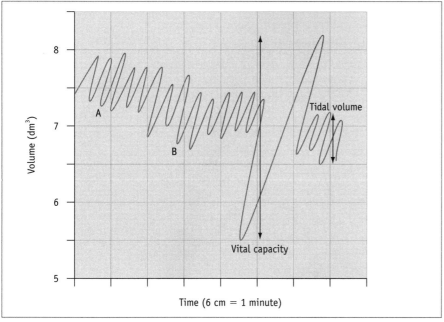

▲ **Figure 8.36** A spirometer trace. The trace shows quiet breathing with one maximum breath in and out. The trace can be used to measure depth and frequency of breathing. The fall in the trace is due to the consumption of oxygen by the subject and the rate of oxygen consumption can be calculated by dividing the decrease in volume by time for the fall. Try analysing this trace by answering Question 8.20.

Topic 8

The volume of air taken into the lungs in one minute, the minute **ventilation**, is calculated by multiplying the tidal volume (the average volume of one breath, in dm^3) by the breathing rate (number of breaths per minute).

Q8.19 At rest, the average person takes 12 breaths per minute. Assuming a tidal volume of 0.5 dm^3, calculate the volume of air breathed in each minute.

Q8.20 From the spirometer trace in Figure 8.36 a) read the tidal volume b) read the vital capacity, c) work out the minute ventilation, and d) work out the volume of oxygen consumed between points A and B on the trace and calculate the rate of oxygen consumption.

The control of breathing

The **ventilation centre** in the medulla oblongata of the brain controls breathing; this is summarised in Figure 8.37. The ventilation centre sends nerve impulses every 2–3 seconds to the external intercostal muscles and diaphragm muscles Both these sets of muscles contract causing inhalation. As the lungs inflate, stretch receptors in the bronchioles are stimulated. The stretch receptors send inhibitory impulses back to the ventilation centre. As a consequence, impulses to the muscles stop and the muscles relax, stopping inhalation and allowing exhalation. Exhalation occurs by the elastic recoil of the lungs (like a deflating balloon), and by gravity helping to lower the ribs. Not all of the air in the lungs is exhaled with each breath, the air remaining in the lungs, residual air, mixes with the air inhaled with each breath.

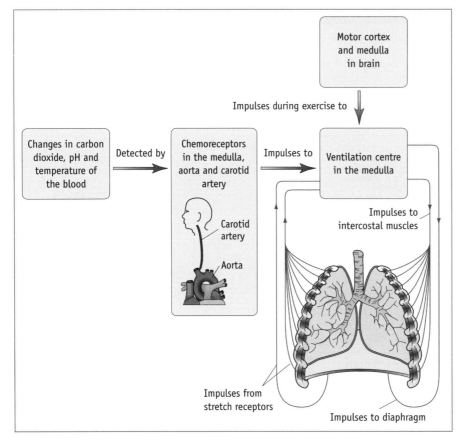

▲ **Figure 8.37** Control of breathing.

The internal intercostal muscles only contract during deep exhalation; for example during vigorous exercise a larger volume of air is exhaled, leaving less residual air in the lungs. During deep inhalation, not only are the external intercostals and diaphragm muscles stimulated but the neck and upper chest muscles are also brought into play.

At rest, the most important stimulus controlling the breathing rate and depth of breathing is the concentration of dissolved carbon dioxide in arterial blood via its effect on pH. A small increase in blood carbon dioxide concentration causes a large increase in ventilation. The carbon dioxide dissolves in the blood plasma, making carbonic acid. Carbonic acid dissociates into hydrogen ions and hydrogen carbonate ions, thereby lowering the pH of the blood:

$$CO_2 + H_2O \rightleftharpoons H_2CO_3 \rightleftharpoons H^+ + HCO_3^-$$

Chemoreceptors sensitive to hydrogen ions (more abundant at low pH) are located in the ventilation centre of the medulla oblongata. Increasing carbon dioxide and the associated fall in pH leads to an increase in rate and depth of breathing, through more frequent and stronger contraction of the appropriate muscles. The more frequent and deeper breaths, maintaining a steep concentration gradient of carbon dioxide between the alveolar air and the blood (Figure 8.38), ensure efficient removal of carbon dioxide and uptake of oxygen. The opposite response occurs with a decrease in carbon dioxide. The control of carbon dioxide levels in the blood is an example of homeostasis operating via negative feedback.

There are also chemoreceptors in the carotid artery and aorta which are stimulated by changes in pH resulting from changes in carbon dioxide concentration. These chemoreceptors monitor the blood before it reaches the brain and send impulses to the ventilation centre.

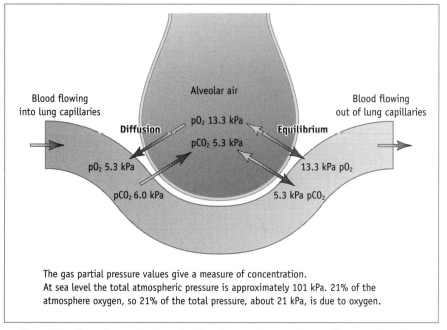

The gas partial pressure values give a measure of concentration.
At sea level the total atmospheric pressure is approximately 101 kPa. 21% of the atmosphere oxygen, so 21% of the total pressure, about 21 kPa, is due to oxygen.

▲ **Figure 8.38** The exchange of carbon dioxide and oxygen between lung capillaries and alveoli.

Topic 8

Q8.21 Describe the sequence of events that will change the rate of breathing if a decline in carbon dioxide occurred in the blood.

Immediately exercise begins, the impulses from the motor cortex of the brain, the region that controls movement, have a direct effect on the ventilation centre, increasing ventilation sharply. Ventilation is also increased in response to impulses reaching the medulla oblongata from stretch receptors in tendons and muscles due to movement. The various chemoreceptors sensitive to carbon dioxide levels and to changes in blood temperature increase the depth and rate of breathing via the ventilation centre. There are receptors sensitive to changing oxygen concentrations in the blood; however they are rarely stimulated under normal circumstances.

Activity

In **Activity 8.11** you can investigate the control of ventilation rate. **A208ACT11**

Q8.22 During vigorous exercise, the concentration of oxygen in the lungs is higher than when at rest. Suggest the reasons for, and the advantage of, this elevated oxygen level.

Q8.23 Suggest why it is beneficial that stimulation of stretch receptors in the muscles increases ventilation.

Q8.24 When a person breathes air containing 80% oxygen, the minute ventilation is reduced by 20%. Explain how this occurs.

Q8.25 Explain why hyperventilation (breathing heavily for several seconds) allows you to hold your breath for longer.

Nice to know: Altitude sickness

The control of breathing can go wrong if we climb too quickly to high altitude. As oxygen becomes scarce (at about 4000 m), blood oxygen concentration can decrease to a very low level. This triggers the medulla oblongata to make us breathe deeply and rapidly, especially if we are doing a hard climb. The hard panting flushes out too much carbon dioxide, resulting in a rise of blood pH. The ventilation centre may respond by stopping breathing altogether for a few seconds, so we alternate between heavy panting and not breathing at all. If this happens it is essential to descend quickly to a lower altitude.

Weblink

You can access data obtained on an Everest climb.

All muscle fibres are not the same

Aerobic capacity is not only dependent on uptake and transport of oxygen to the muscles, but also on efficiency of use once it reaches the muscle fibres. Although the muscles of all mammals are almost identical in their macroscopic and microscopic structure, it is possible to identify two distinct types of fibre.

If you eat chicken (which has the same types of muscle fibres as mammals) you may have noticed that some parts of the flesh are darker and others lighter. The breast meat, where the flight muscles are, is pale (the 'white meat') whereas the leg muscles are darker. This difference is caused by the fact that the two regions contain two different types of muscle fibre, which reflects the different functions of the muscles.

Chickens and turkeys spend most of their time on the ground, standing still or walking. However, in moments of panic or to reach higher perches they can perform near vertical take-offs, though they cannot fly for long.

The darker muscle in the legs and body is made up of fibres called **slow twitch fibres**. These are specialised for slower, sustained contraction, and can cope with long periods of exercise. To do this they need to carry out a large amount of aerobic respiration.

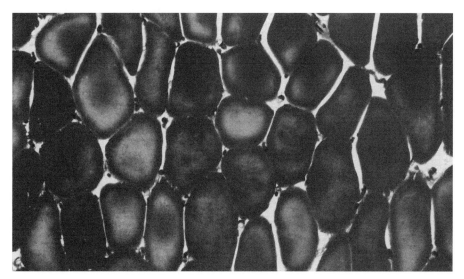

▲ **Figure 8.39** Notice how some of the fibres are very dark. These contain more myoglobin and are slow twitch.

The slow twitch muscle fibres have many mitochondria and high concentrations of respiratory enzymes to carry out the aerobic reactions. They also contain large amounts of the dark red pigment **myoglobin**, which gives them their distinctive colour. Myoglobin is a protein similar to haemoglobin (the oxygen-carrying pigment found in red blood cells). It has a high affinity for oxygen and only releases it when the concentration of oxygen in the cell falls very low; it therefore acts as an oxygen store within muscle cells. Slow twitch fibres are associated with numerous capillaries to ensure a good oxygen supply.

The paler flight muscle is largely made up of a different type of muscle fibre called **fast twitch fibres**. These fibres are specialised to produce rapid, intense contractions. The ATP used in these contractions is produced almost entirely from anaerobic glycolysis.

The fast twitch fibres have few mitochondria. (Remember that glycolysis does not occur in mitochondria.) They also have very little myoglobin, so have few reserves of oxygen. With a rapid build-up of lactate, the fast twitch muscle fibres fatigue easily. With aerobic training fast twitch fibres can take on some of the characteristics of slow twitch fibres, for example increased numbers of mitochondria, allowing them to use aerobic respiration reactions when contracting.

Q8.26 Study Table 8.2 and then answer the following questions.
a) What is the significance of the large number of mitochondria in slow twitch fibres?
b) Why do fast twitch fibres need more sarcoplasmic reticulum?
c) Why will a large amount of myoglobin be advantageous to the slow fibres?
d) How will each type of fibre regenerate ATP?
e) Which type of fibre will build up an oxygen debt more quickly? Give a reason for your answer.

▼ Table 8.2 Characteristics of the two muscle types.

Slow twitch fibres	Fast twitch fibres
Red (a lot of myoglobin)	White (little myoglobin)
Many mitochondria	Few mitochondria
Little sarcoplasmic reticulum	Extensive sarcoplasmic reticulum
Low glycogen content	High glycogen content
Numerous capillaries	Few capillaries
Fatigue-resistant	Fatigue quickly

In mammals, these two types of muscle fibre are not separated but found together in all the skeletal muscles. The proportion of each type appears to be genetically determined, and varies between different people. Research has shown that successful endurance athletes such as marathon runners, rowers and cross-country skiers, with high aerobic capacity, have a higher proportion of slow twitch muscle fibres, up to 80% in their skeletal muscles. In contrast, sprinters may have as few as 35% slow twitch fibres. Throwers and jumpers have a more-or-less equal proportion of the two types.

Individuals may be better suited to a particular type of sport if they naturally have a higher proportion of fibres used in that activity. This, of course, will not be the only factor that contributes to an individual's success in a sport. For example, an individual with a highly efficient cardiovascular system will be well suited to aerobic exercise.

Q8.27 Which type of fibre might dominate the leg muscles of a) the cheetah and b) the wildebeest?

8.4 Breaking out in a sweat

The demands of physical activity can increase metabolic activity by up to 10 times resulting in a heat production which could potentially increase core temperature by around 1 °C every 5 to 10 minutes. This energy must be dispersed to maintain thermal balance.

The marathon, more than any other sport, has a history of heat-related deaths. In 490 BC Pheidippides, an intercity messenger of Ancient Greece, ran 26 miles from Marathon to Athens with instructions that the Athenians should not surrender to the Persian fleet. Legend has it that at the end of his journey Pheidippides dropped dead from exhaustion. In the 1912 Olympic Games in Stockholm, the Portuguese runner Lazaro collapsed from heat stroke after running 19 miles. He died the following day. Nowadays marathons tend to be scheduled in the early morning to reduce the chance of heat stroke. However, this is unpopular with TV schedulers and runners are sometimes required to race in more dangerous conditions to maximise advertising revenues.

▲ **Figure 8.40** During strenuous exercise enough heat is produced to raise our body temperature by 1 °C every 5 to 10 minutes were it not dissipated.

Q8.28 a) What is the optimal temperature range for human cells?
b) What will happen to metabolic reactions if body temperature sinks below or rises above the normal range?

As we saw in Topic 7, core temperature is normally maintained within the narrow range of 36–38 °C in humans using negative feedback. A rise of only 5 °C in core body temperature can be fatal.

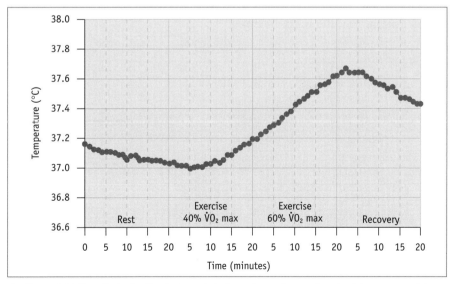

▲ **Figure 8.41** The effect of a 40-minute period of exercise on temperature (rectal measurements).

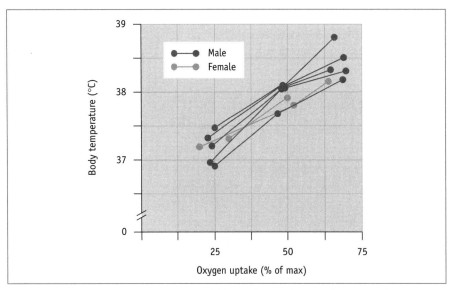

▲ **Figure 8.42** Changes in body temperature with increasing exercise intensity expressed as oxygen uptake.

During exercise, temperature rises (Figure 8.41) and this rise is related to intensity of exercise (Figure 8.42). The comparatively slight rise in temperature does not indicate a failure of the body to regulate the temperature; without mechanisms to redistribute energy though, the increase would quickly reach dangerous levels.

Thermoregulation during exercise is a good example of a homeostatic process; the body maintains a constant optimal internal temperature using negative feedback. The body must:

- detect changes in internal conditions using receptors
- coordinate the action of effectors to oppose the change and reset the normal conditions.

Once the hypothalamus detects a deviation from the norm in core temperature, it starts a chain of reactions which will counteract the deviation and bring body temperature back to the norm.

Use Figure 8.43 to remind yourself of the ways you can raise and lower body temperature.

 Activity

Use **Activity 8.12** to remind yourself how negative feedback achieves temperature regulation. **A208ACT12**

Run for your life

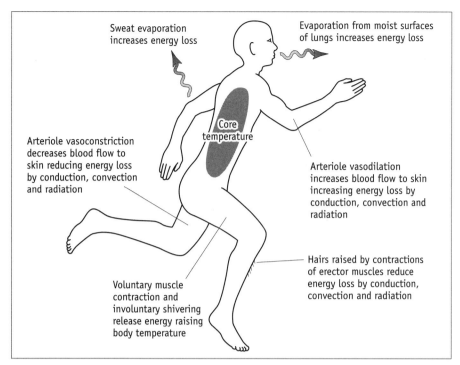

▲ Figure 8.43 How energy is transferred to and from the body.

Methods of energy transfer

Radiation – Energy can be radiated from one object to another through air, or even a vacuum, as electromagnetic radiation. Our bodies are usually warmer than the surrounding environment, so we radiate energy. Of course, this can operate in reverse – even in sub-freezing conditions such as skiing in the mountains, a person can remain warm due to the heat energy radiated directly from the Sun, or reflected back from the snow.

Conduction – Energy loss by conduction involves direct contact between objects and energy transfer from one to another. Sitting on a cold rock on a hot day will cause body cooling, as energy is transferred to the rock by conduction.

Convection – Air lying next to the skin will be warmed by the body (unless air temperature exceeds body temperature). If the air expands and rises, or is moved away by air currents, it will be replaced by cooler air which can then also be warmed by the body. This is energy loss by convection. Trapping a layer of still air next to the skin using fur or thermal underwear is an effective method of thermal insulation.

Evaporation – Energy is needed (latent heat of evaporation) to convert water from liquid to vapour. 1 dm^3 of water requires 2400 kJ to evaporate. The energy required to evaporate sweat is drawn from the body, cooling it. However, sweating is only effective if the water actually evaporates from the surface of the skin. In conditions of high humidity this becomes virtually impossible, and controlling body temperature becomes much more difficult. Mammals, birds and reptiles may also pant to keep them cool, with evaporation of water from the gas exchange surfaces.

Q8.29 The marathon runner in Figure 8.40 feels cooler after dousing himself in water. How does this help to cool him?

Topic 8

The maximum distance a cheetah can sprint is approximately 500 metres. After this point, not only will the fast twitch muscle fibres have fatigued due to the build-up of lactate, but the body temperature will have risen. Cheetahs must stop to recover.

Q8.30 A cat (including the big cats) can only sweat from the skin surface of its paws and nose. What other method might the cheetah rely on for transfer of energy to the environment?

At 37 °C, human core body temperature is normally higher than the surroundings, so energy will be transferred to the environment. In very cold environments, excessive cooling may occur and the body can lose thermal balance – the core body temperature starts to fall. The hypothalamus detects this and immediately does its best to regulate the internal temperature by increasing metabolic rate and slowing energy loss, as outlined in Topic 7. Although less common during exercise, there are occasions when the body faces this challenge (see Figure 8.44).

▲ **Figure 8.44** Channel swimming exposes the body to cold stress.

Q8.31 a) What will be the major method of energy loss for the cross-Channel swimmer in Figure 8.44?
b) Swimming the English Channel in summer means spending between 10 and 20 hours in water that is usually between 13 and 16 °C. What mechanisms are used to maintain body temperature and enable survival without wearing a wetsuit?

8.5 Overdoing it

▲ **Figure 8.45** Top sportspeople can suffer from overtraining. At the end of the season some professional footballers have been found to have very low white blood cell counts.

Run for your life

Although we will never run as fast as a sprinting cheetah, over the years top athletes have been getting faster. At the end of 1920s the 100 m world records stood at 10.4 and 12.0 seconds for men and women respectively. Since then, the times have steadily fallen, and at the dawn of the new millennium the records stood at 9.79 and 10.49 seconds. This improvement in performance has been achieved through more frequent and targeted training, improved nutrition, and advances in the design and materials used in athletes' clothing, footwear and tracks.

But some athletes attempt to do more than their body can physically tolerate and reach the point where they have inadequate rest to allow for recovery (Figure 8.45). This is known as overtraining. 'Burnout' symptoms due to overtraining can persist for weeks or months. The symptoms are varied and, in addition to poor athletic performance and chronic fatigue, can include immune suppression leading to more frequent infections and increased wear and tear on joints, which may require surgical repair.

Why can excessive exercise lead to immune suppression?

Athletes engaged in heavy training programmes seem more prone to infection than normal. Sore throats and flu-like symptoms (**upper respiratory tract infections**, **URTIs**) are more common. Some scientists have suggested there is a U-shaped relationship between risk of infection and amount of exercise (Figure 8.46).

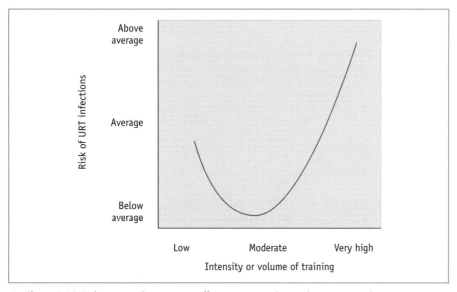

▲ **Figure 8.46** Moderate exercise seems to offer some protection against upper respiratory tract infections.

Q8.32 What do these data suggest about the risk of upper respiratory tract infections related to exercise?

Topic 8

In a study of participants in the Los Angeles marathon, it was found that 13% of the runners reported upper respiratory tract infections in the week after the race. Runners of comparable fitness who had not been able to race for reasons other than illness had an infection rate in the week after the race of only 2%. There is much discussion about whether this is a true cause and effect relationship.

Two main factors have been suggested as contributing to higher infection rates: increased exposure to pathogens, and suppressed immunity with hard exercise.

The location of the competition and any necessary travel may expose the athlete to a greater range of infected people and unfamiliar microorganisms. This alone could increase the occurrence of infection, even if overtraining did not suppress their immunity. Participation in team sports will also bring players into close contact with others and increase the chances of transmission of infection.

A number of research studies have shown that different components of both the non-specific and specific immune systems are affected to various degrees by both moderate and excessive exercise. During moderate exercise the number of **natural killer cells** increases, as does their activity.

Natural killer cells are a group of lymphocytes found in the blood and lymph; unlike the B and T cells, they do not use specific antigen recognition. They provide non-specific immunity against viruses and other intracellular microbes or cancerous cells. They recognise sugars on the surface of the target cell and secrete apoptosis-inducing molecules, causing lysis of the cells. They thus offer non-specific protection against upper respiratory tract infections and other infections.

Research shows that during recovery after vigorous exercise, the number and activity of some cells in the immune system fall. These include:

- natural killer cells
- phagocytes
- lymphocytes
- helper T cells.

The decrease in helper T cells reduces the amounts of cytokines available to activate T and B cells. This in turn reduces the quantity of antibodies produced. It has also been suggested that an inflammatory response occurs in muscles due to damage to muscle fibres caused by heavy exercise, and this may reduce the available non-specific immune response against upper respiratory tract infections.

Q8.33 a) Which of the white blood cells mentioned above secretes antibodies?
b) What are the main features of the non-specific immune response to infection?
c) How will the action of killer T cells be affected by the decrease in helper T cell numbers?

There is also much debate as to whether the effects of intense exercise are caused by the activity itself or by related psychological stress due to heavy training schedules and competition. Both physical exercise and psychological stress cause secretion of hormones such as adrenaline and cortisol (a hormone also secreted by the adrenal glands), both of which are known to suppress the immune system.

Q8.34 How does the evidence above support the idea that moderate exercise enhances immunity while excessive exercise suppresses it?

Activity

In **Activity 8.13** you can summarise your knowledge of the immune system and immune suppression. **A208ACT13**

How are joints damaged by exercise?

Professional athletes, such as football, hockey and rugby players, risk developing joint injuries due to the high forces the sport generates on their joints. Repeated forces on joints such as the knee can lead to wear and tear of one or more parts of the joint. A number of joint disorders are associated with such overuse, many of which can also result from ageing. These disorders are typically associated with pain, inflammation and restricted movement of the joint. Treatment usually involves rest, ice, compression and elevation (RICE), anti-inflammatory painkillers, and, if necessary, surgical repair.

Knees are particularly susceptible to wear and tear injuries. The problems include:

- The articular cartilage covering the surfaces of the bones wears away so that the bones may actually grind on each other, causing damage which can lead to inflammation and a form of arthritis.

- Patellar tendonitis (jumper's knee) occurs when the kneecap (patella) does not glide smoothly across the femur due to damage of the articular cartilage on the femur.

- The bursae (fluid sacs) which cushion the points of contact between bones, tendons and ligaments can swell up with extra fluid. As a result, they may push against other tissues in the joint, causing inflammation and tenderness. Bursitis of the knee is also known as 'housemaid's knee' because it was common in housemaids due to the repetitive bending associated with their work.

- Sudden twisting or abrupt movements of the knee joint often result in damage to the ligaments.

Topic 8

How can medical technology help?

Improvements in medical technology over recent years, including the development of prosthetic limbs and keyhole surgery procedures, have enabled the disabled and those with injuries to participate in sport.

Keyhole surgery

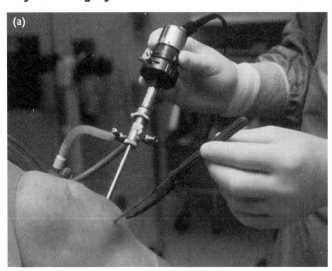 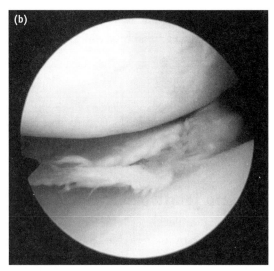

▲ **Figure 8.47** Arthroscopy (a) allows surgeons to see within joints and repair damage like the degenerative tear in this knee joint (b).

Injuries to joints have often shortened the careers of professional athletes. Surgical operations to repair damage used to be painful and recovery took a long time. The main reason for this was the large incision needed to remove or repair even very small structures. A large hole had to be made to give access with space for the surgeon's hands and instruments, and also let in enough light to allow the surgeon to see what she/he was doing. These large incisions caused a good deal of bleeding, a lot of pain, increased risk of infection, and prolonged recovery after the operation.

With the advent of **keyhole surgery**, using fibre optics or minute video cameras, all this has changed. It is now possible to repair damaged joints or remove diseased organs through small holes. Keyhole surgery on joints is known as **arthroscopy** (Figure 8.47a). This literally means "to look within the joint". To carry out an arthroscopic examination, the surgeon makes only a very small incision into the skin of the patient. Then small surgical instruments about the size of a pencil and a fibre optic tube to view the inside of the joint are placed inside the joint. A TV camera and light are attached to the fibre optic tube, to allow the surgeon to see inside the joint (Figure 8.47b) through this small incision. As a result of the small incision, recovery is rapid, and only a short stay in hospital is needed.

Although the inside of most joints can be viewed with an arthroscope, the six joints most frequently examined and treated are the knee, shoulder, elbow, ankle, hip and wrist. As advances are made in medical technology and new techniques are developed by surgeons, other joints may be treated more frequently in the future.

Prostheses

▲ **Figure 8.48** Two different types of leg prostheses in use by athletes.

A **prosthesis** (plural prostheses) is an external artificial body part used by someone with a disability to enable him or her to regain some degree of normal function or appearance. By using specialized prostheses, disabled athletes can be more physically active and perform at higher levels (Figure 8.48).

There have been significant developments in prosthetic limbs over recent years, with the introduction of variations in design for different activities. For example, athletes with prosthetic legs may use a dynamic response prosthetic foot. Such a foot changes its shape under body weight but returns to the original shape on lifting off the ground. This puts spring into the step and provides sure footing.

Prosthetic feet may also be articulated (have joints) or not. Articulated feet are better on uneven surfaces, and so useful in sports such as golf. In some sports, a foot may not be required at all; a flipper could be used for swimming, or a pedal-binding for cycling. High friction surfaces can be added to provide better grip to prostheses used for rock climbing.

> **Activity**
>
> Check out the video on bone damage and repair in **Activity 8.14. A208ACT14**

8.6 Improving on nature

Performance-enhancing substances

The use of drugs to enhance performance in sport is known as **doping**. It is thought that this term originates from the South African word 'doop', which referred to an alcoholic stimulant drink used in certain tribal ceremonies.

Doping is not just a recent problem; throughout history, some athletes have sought a competitive edge by the use of chemicals. As long ago as the 3rd century BC, certain Ancient Greeks were known to ingest hallucinogenic mushrooms to improve athletic performance. Roman gladiators used stimulants in the Circus Maximus to overcome fatigue and injury, while other athletes experimented with caffeine, alcohol and opium. The important social status of sport, and the high economic value of victory, both then and now, has placed great pressure on athletes to be the best. This pressure has escalated the abuse of performance-enhancing drugs.

▲ **Figure 8.49** Ancient drug cheats?

A wide range of substances can be taken for their performance-enhancing effects. Three examples are **erythropoetin**, **testosterone** and **creatine**. Erythropoetin and testosterone use are currently banned whereas creatine is not.

Nice to know: Banned substances

WADA, the World Anti-Doping Agency, promotes and coordinates the international fight against doping in sport. WADA aims to reinforce the "ethical principles for the practice of doping-free sport and to help protect the health of athletes".

The Olympic Movement Anti-Doping Code prohibits various substances and practices. The following classes of substances are prohibited:

A: stimulants (increase heart rate and alertness) such as amphetamines, cocaine and excessive amounts of pseudoephedrine (a drug commonly found in over-the-counter decongestants) and caffeine (found in coffee, tea and chocolate).

B: narcotics (powerful painkillers causing drowsiness) such as diamorphine (heroin), methadone, morphine and pethidine.

C: anabolic agents. This class includes anabolic steroids, such as testosterone and nandrolone, and beta–2 agonists such as salbutamol, a bronco-dilator used in asthma inhalers.

D: diuretics (drugs that promote formation of urine) such as frusemide.

E: peptide hormones, mimetics and analogues, including EPO (erythropoetin) and insulin.

Competition athletes with medical conditions requiring prescription of these drugs need to obtain permission to use them. It is the responsibility of the athlete to check whether a drug is banned or not, and to ensure that he/she does not inadvertently take the banned substance (Figure 8.50).

▲ **Figure 8.50** Alain Baxter, the British skier, was stripped of his bronze medal at the 2002 Winter Olympics for testing positive for a banned substance after using a nasal inhaler.

Topic 8

Erythropoetin

Erythropoetin (EPO) is a peptide hormone produced naturally by the kidney. It stimulates the formation of new red blood cells in bone marrow. EPO can be produced using DNA technology, and is used to treat anaemia. As it is a natural substance, it has been difficult to test to see whether raised EPO levels are natural or not.

Q8.35 Explain how taking EPO would increase the performance of an endurance athlete.

There are health risks associated with this substance. If EPO levels are too high, the body will produce too many red blood cells, which can increase the risk of thrombosis, possibly leading to heart attack and stroke. Injections of EPO have been implicated in the deaths of several athletes.

French scientists have recently developed a technique capable of distinguishing between synthetic and natural EPO.

Q8.36 What is meant by thrombosis?

Q8.37 Why is EPO not taken by sprint athletes?

Q8.38 Why would it be important to distinguish between natural and recombinant EPO?

Testosterone

Testosterone is a steroid hormone (made from cholesterol) produced in the testes by males and in small amounts by the adrenal glands in both males and females. Testosterone is one of a group of male hormones known as androgens, from the Greek *andros* meaning male or man.

Testosterone causes the development of the male sexual organs. During adolescence it is responsible for development of the male secondary sexual characteristics, for example the deepening of the voice, growth of facial and body hair, and skeletal and muscular changes. Character changes such as increased aggressiveness have been attributed to testosterone.

Testosterone binds to androgen receptors which are numerous on cells in target tissues. They modify gene expression to alter the development of the cell; for example they will increase anabolic reactions such as protein synthesis in muscle cells, increasing the size and strength of the muscle.

Athletes and body builders (Figure 8.51) may use injections of testosterone to increase muscle development, but this is not very effective as testosterone is quickly broken down. To overcome this problem, synthetic **anabolic steroids** such as nandrolone have been manufactured by chemical modification of testosterone.

Medical experts see significant dangers in the use – and particularly the gross over-use – of anabolic steroids. For example, anabolic steroids can cause high blood pressure, liver damage, changes in the menstrual cycle in

Activity

In **Activity 8.15** you interpret data on the effects of testosterone.
A208ACT15

▲ **Figure 8.51** Anabolic steroids are used to increase muscle development. Heavy-weight or resistance training is necessary for anabolic steroids to exert any beneficial effect on performance.

women, decreased sperm production and impotence in men, kidney failure and heart disease. They can increase aggression in both men and women. In women the androgenic (masculinising) side-effects are not generally thought to be desirable.

Originally developed for the treatment of muscle-wasting diseases, anabolic steroids are also used in the treatment of osteoporsis. In the UK they are prescription-only drugs. They are classified as Class C drugs under the Misuse of Drugs Act, with the maximum penalty for the illegal *possession* of steroids currently (2003) standing at two years imprisonment and/or a fine. *Supplying* a Class C drug, such as an anabolic steroid, can lead to heavier penalties, even if no money has changed hands. The International Olympic Committee has banned the use of anabolic steroids. The illegal use of steroids not only occurs in human sport but also animal sports such as horse racing and dog racing.

Anabolic steroids and their by-products can be detected relatively easily in urine samples by the technique of mass spectrometry. As these substances occur naturally it is difficult to set a level above which an athlete can confidently be said to be doping. Testosterone and a related compound, epitestosterone, are both found in urine. When an athlete takes an anabolic steroid, the ratio of testosterone to epitestosterone (the T/E ratio) increases. The International Olympic Committee states that an athlete with a T/E ratio above 6 is guilty of doping.

Creatine

Many athletes take dietary supplements containing an amino-acid derived compound known as creatine. Creatine is naturally found in meat and fish. Once ingested it is absorbed unchanged and carried in the blood to tissues such as skeletal muscle. Creatine is also synthesized in the body from the amino acids glycine and arginine. Creatine supplements have been reported to increase the amounts of creatine phosphate (CP) in muscles. The theoretical benefit of increased CP storage is an improvement in performance during repeated short-duration, high-intensity exercise. Research has shown improvements in activities such as sprinting, swimming and rowing. The use of creatine supplements combined with heavy weight training has been associated with increases in muscle mass and maximal strength, and a decrease in recovery time.

Creatine is considered as a nutritional supplement and is therefore not on the list of prohibited substances and so its use is not banned. Some adverse effects of taking creatine supplements have been reported. These include diarrhoea, nausea, vomiting, high blood pressure, kidney damage and muscle cramps.

Topic 8

Key biological principle: Hormones

Hormones are chemical messengers released from **endocrine glands** directly into the blood. (Unlike exocrine glands, such as sweat glands and salivary glands, endocrine glands do not have ducts). They are carried around the body and enter cells or bind to complementary receptor molecules on the cell membranes of specific target cells, as described in Topic 3. Hormones affect enzymes which leads to the characteristic response.

Peptide hormones are protein chains, varying from about 10 to 300 amino acids in length. Peptide hormones cannot easily pass through a cell membrane because they are charged molecules. The binding of a peptide hormone to a receptor in the cell membrane activates another molecule in the cell cytoplasm which brings about chemical changes in the cell. Peptide hormones include EPO, human growth hormone and insulin.

Steroid hormones are formed from lipids, and have complex ring structures. Testosterone is a steroid hormone. Steroid hormones pass through the cell membrane and bind to receptor molecules within the cell which, once activated, function as transcription factors.

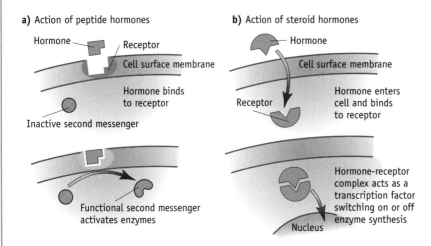

▲ **Figure 8.53** Hormones affect target cells by binding to receptors which control production of enzymes.

Q8.39 Why can steroid hormones pass through the cell membrane?

Should performance-enhancing substance use be banned?

The pressure to succeed in competitive sport is every increasing, not only due to the desire to be the best but also for the financial rewards and greater media interest. The desire to win combined with pressure and expectations from coaches, sponsors and the general public can be such that some athletes are prepared to take drugs that will enhance their performance, even if there are associated risks.

Activity

Think about the ethical arguments about doping in **Activity 8.16**.
A208ACT16

The International Olympic Committee (IOC) and other sporting bodies consider that the use of performance-enhancing substances is unhealthy and against the ethics of sport. The IOC started drug testing in 1968 after a Tour de France cyclist died from an amphetamine overdose; random testing began in 1989. The ban aims to protect the health of athletes and ensure that there is fair competition.

There are some people who consider that the use of substances is ethically acceptable, arguing that athletes have a right to decide whether they take the drug or not, deciding for themselves if the potential benefit is worth the risk to their health. Those who oppose this view may say that frequently the athletes do not make a properly informed decision, lacking information about the possible health consequences, and coming under pressure from others to take illegal drugs.

The idea that drug-free sport is fair is disputed by those who maintain that drug use is acceptable on the grounds that there is already inequality of competition due to the differences in time available for training and in resources. Some individuals may know that drug use is against the rules both of the governing bodies of sport and the idea of fair play but they are unwilling to be at a competitive disadvantage so choose not to adhere to the rules.

Summary

Having completed Topic 8 you should be able to:

- Recall the way in which muscles, tendons, the skeleton and ligaments interact to enable movement.

- Explain the contraction of skeletal muscle in terms of the sliding filament theory (including the role of actin, myosin, troponin, tropomyosin, Ca^{2+}, ATP).

- Explain how ATP is the immediate supply of energy for biological processes.

- Describe the roles of glycolysis in aerobic and anaerobic respiration, starting with phosphorylation and ending with pyruvate (names of other compounds are not required).

- Describe the role of the Krebs cycle in the complete oxidation of glucose and formation of CO_2, ATP, reduced NAD and reduced FAD (names of other compounds are not required).

- Describe the synthesis of ATP associated with the electron transport chain in mitochondria.

- Explain the fate of lactate after a period of anaerobic respiration.

- Discuss why some mammals are better at short bursts of high intensity exercise while others are better at long periods of continuous activity.

- Describe the structural, and explain the physiological, differences between fast and slow twitch muscle fibres.

Topic 8

- Explain how variations in ventilation and cardiac output enable efficient delivery of oxygen to tissues and removal of carbon dioxide from them.
- Describe how to investigate the effects of exercise on tidal volume and breathing rate.
- Discuss the concept of homeostasis and its importance in maintaining the body in a state of dynamic equilibrium during exercise as exemplified by thermoregulation.
- Discuss possible disadvantages of exercising too much (wear and tear on joints, suppression of the immune system).
- Explain how medical technology, including the use of key-hole surgery and prostheses, is enabling those with injuries and disabilities to participate in sports.
- Discuss whether the use by athletes of performance-enhancing substances, including creatine, testosterone and erythropoetin, is morally and ethically acceptable.

 Review test

Now that you have finished Topic 8, complete the end-of-topic test before starting Topic 9. **A208RVT01**

Grey matter

Topic 9

Grey matter

Why a topic called *Grey matter*?

The brain, with over 10^{12} neurones (nerve cells), is the most complicated organ in the body and makes even our most sophisticated computer seem simple. The brain influences our every sensation, emotion, thought, memory and action. At each moment of every day it is bombarded with sensory information from the world around us, and interprets this information to create a meaningful view of the world. Looking at the world is not merely inspecting a simple picture, like observing a slide show projected on a screen inside our heads. The information is processed to provide us with our experience of the world.

But sometimes things may not be as they seem, as the anthropologist Colin Turnbull found in the 1950s. He and Kenge, one of the Bambuti people used to living in dense forest with only small clearings, went out onto the grassland plains of the former Congo. They saw buffalo grazing several miles away (Figure 9.1). Kenge turned to Turnbull and asked 'What insects are those?'. When Turnbull told Kenge that the insects were buffalo, he roared with laughter and told him not to tell such stupid lies.

▲ **Figure 9.1** Kenge thought the buffalo in the distance were insects.

How does the nervous system function to let any of us look across a plain? Why did Kenge get the wrong impression? Was his visual development faulty or was he misinterpreting what he was viewing?

It is not just Kenge who is sometimes mistaken by what is viewed. Have a look at Figure 9.2 and decide which of the lines is longer. Now measure them to find out if you were correct. Why do many of us get this wrong when Zulu people are not fooled so easily?

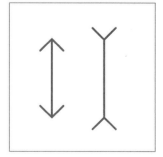

▲ **Figure 9.2** Which line is longer?

All in the synapses

With upwards of 10^{14} interconnections between its neurones, the working of the brain is dependent on its synapses and their neurotransmitters. How do these function and how can they go wrong? Imbalances in naturally-occurring brain chemicals and drugs that can cross the blood-brain barrier can affect synapses and have adverse consequences for health. How are synapses affected by conditions such as depression and Parkinson's disease and by the use of MDMA (ecstasy)?

Overview of the biological principles covered in this topic

In this topic, you will revisit ideas about receptors and effectors introduced in Topics 7 and 8 by examining the pupil light reflex. Building on these ideas you will consider the detection of stimuli in greater detail, as exemplified by light detection by receptor cells in the retina, and how the brain and eyes combine to enable visual perception. This leads to discussion of the transmission of nerve impulses along axons and across synapses before contrasting nervous and hormonal coordination.

You will look at how some diseases and drugs affect the brain to illustrate how chemicals can affect synaptic transmission. You will revisit genetic inheritance studied in Topics 2 and 6 and see how many mental disorders show polygenic inheritance. With reference to some of the DNA techniques encountered earlier in the course you will discover how genetics can play a part in increasing our knowledge of brain structure and activity.

You will investigate the structures of the different regions of the brain and the evidence that links structure and function including the use of imaging techniques you met in Topic 1. You will look at visual development and in particular the need for stimulation of synapses and the role of synapses in learning. Throughout this topic the contribution of nature and nurture to brain development is highlighted.

You will have the opportunity to discuss the ethical issues related to the use of animals in research.

Topic 9

9.1 How did Kenge see the image?

As Kenge and Colin Turnbull emerged from the forest, how did their eyes and brains work to let them look across the plain? Seeing is possible because the cells of the nervous system (Figure 9.3) are able to conduct **nerve impulses** and pass them to one another. In fact all our senses, emotions, memory and thoughts are dependent on nerve impulses.

The nervous system is highly organised (Figure 9.4.) receiving, processing and sending out information as we saw with temperature control in Topic 7 and control of heart rate in Topic 8.

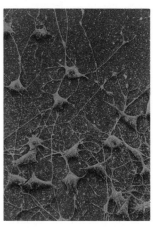

▲ **Figure 9.3** Nerve cells such as these form the basis of the nervous system.

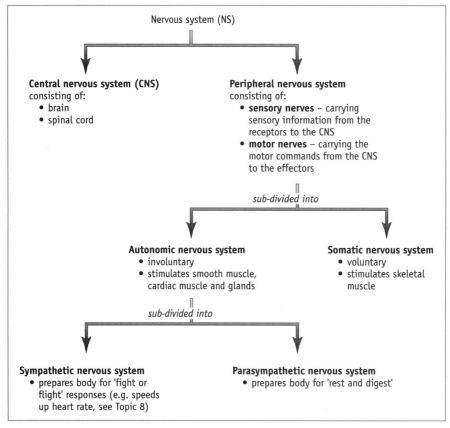

▲ **Figure 9.4** The organisation of the nervous system.

What are nerve cells like?

Although there are different types of neurone, they all have the same basic characteristics, which can be seen in Figure 9.3. The **cell body** contains the nucleus and cell organelles within the cytoplasm. There are two types of thin extensions from the cell body:

- Very fine dendrites conduct impulses towards the cell body.
- A single long process, the **axon**, transmits impulses away from the cell body.

Figure 9.5 shows the structure of a **motor neurone**. The cell body is always situated within the CNS and the axon extends out, conducting impulses from the CNS to effectors, i.e. muscles or glands. The axons of some motor neurones can be extremely long, such as those that run the full length of the leg.

Sensory neurones carry impulses from sensory cells to the CNS. **Interneurones** are found mostly within the CNS. These can have a large number of connections with other nerve cells.

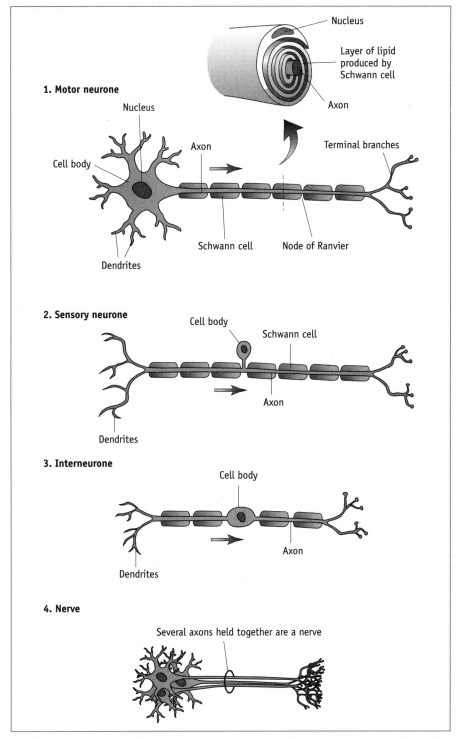

▲ **Figure 9.5** The structure of neurones. The scanning electron micrograph shows the myelin sheath surrounding the axons within a nerve.

Topic 9

There is usually a fatty insulating layer called the **myelin sheath** around the axon. This is made up of **Schwann cells** wrapped around the axon. The sheath affects the speed of conduction (see page 64). Not all animals have myelinated axons – they are not found in invertebrates and some vertebrate axons are unmyelinated.

It is important to distinguish between a neurone, which is a single cell, and a **nerve**. A nerve is a more complex structure containing a bundle of the axons of many neurones surrounded by a protective covering.

Reflex arcs

Nerve impulses follow routes or **pathways** through the nervous system. Most nerve pathways involve numerous neurones within the central nervous system, but some are relatively simple. A **reflex arc** is such a simple pathway. In some cases it involves as few as two neurones when a sensory neurone communicates directly with a motor neurone to connect receptor cells with effector cells for example in the knee jerk reflex. These arcs are responsible for our reflexes, i.e. rapid, involuntary responses to stimuli. Figure 9.6 illustrates one reflex arc.

Q9.1 What is the advantage of reflex pathways?

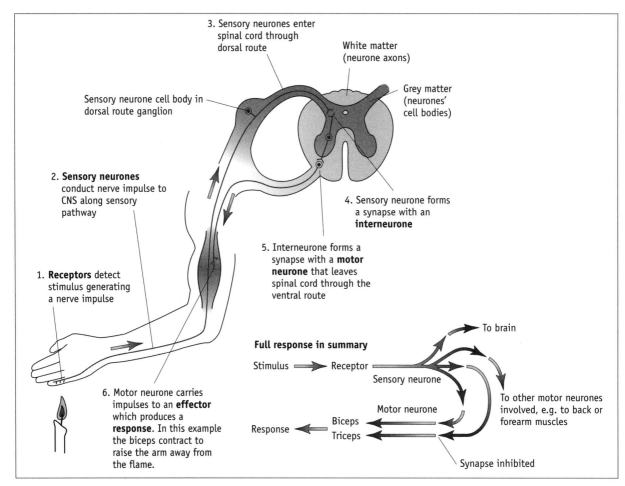

▲ **Figure 9.6** A reflex arc allowing withdrawal of the hand.

Q9.2 It is possible to prevent the withdrawal reflex which occurs, for example, if you touch something hot. What does this suggest about the idea that the pathway is simple?

The pupil reflex

When Kenge and his companion emerged from the trees they moved from deep shade to bright sunlight. Immediately a reflex arc caused a change in the diameter of their pupils. If you cover your eyes for a few minutes, and then uncover your eyes while looking in a mirror, you can see that the size of your pupils decreases and the size of the irises increases (see Figure 9.7).

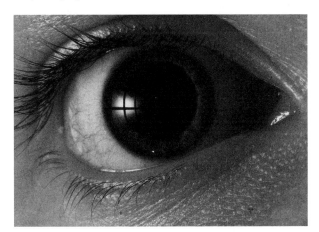
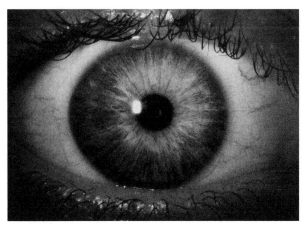

▲ **Figure 9.7** The pupil dilates or constricts in response to changing light intensities.

Q9.3 Which of the eyes in Figure 9.7 is in low light?

How do the muscles of the iris respond to light?

The iris controls the size of the pupil. It contains a pair of antagonistic muscles: **radial** and **circular** muscles (see Figure 9.8). These are both controlled by the autonomic nervous system. The radial muscles are like spokes of a wheel, and are controlled by a sympathetic reflex. The circular muscles are controlled by a parasympathetic reflex. One reflex dilates, the other constricts the pupil.

Q9.4 Which of the two sets of muscles will cause the pupil to dilate?

High light levels striking the photoreceptors in the retina cause nerve impulses to pass along the optic nerve to a number of different sites within the CNS including a group of cells in the mid-brain; these cells control the circular muscles in the iris. Impulses from these coordinator cells are sent along parasympathetic motor neurones to the circular muscles of the iris, causing them to contract. This contracts the size of pupil, reducing the amount of light entering the eye. Figure 9.8 shows the reflex pathway involved.

Topic 9

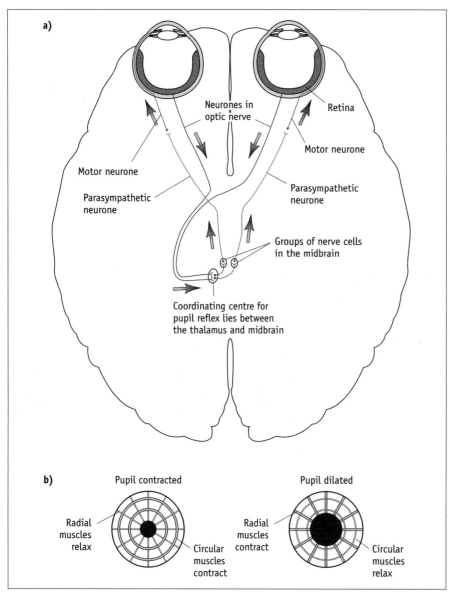

▲ **Figure 9.8** a) The reflex pathway involved in pupil contraction. Pupil dilation involves sympathetic neurones not shown here. b) Muscle action in pupil contraction and dilation.

Q9.5 Name the components *a* to *d* below involved in the pupil reflex.

Receptors	a)
Sensory nerve fibres	b)
Co-ordinator	c)
Motor nerve fibres	Oculomotor nerve
Effector	d)

Q9.6 What is the purpose of the pupil reflex?

Q9.7 The pupil reflex to increased light is very rapid. Why does this need to be the case?

 Activity

Investigate the pupil reflex in **Activity 9.1**. **A209ACT01**

Q9.8 How many synapses are there in the pupil reflex pathway shown in Figure 9.8?

> **Nice to know:** Atropine
>
> Deadly nightshade *(Atropa belladonna)* is the source of the drug atropine. Atropine was used in the Middle Ages by women to make their pupils dilate. This was thought to make them more attractive to men. Atropine inhibits parasympathetic stimulation of the iris. So the circular muscles of the iris relax. Today it is used to dilate the pupils for an eye examination. It is known as an acetylcholine antagonist. When you have completed all the work on the nervous system you should understand what this means!

How do nerve cells transmit impulses?

To understand how Kenge saw, we must first understand how nerve cells transmit impulses, and how the receptor cells in the eye detect light, causing impulses to be sent to the brain where the signals are interpreted.

Much of the work done to establish what happens in a nerve fibre was carried out on the giant axons of the squid (Figure 9.9). Their large size makes them easier to work with. Hodgkin, Huxley and Eccles carried out this work in the 1940s and 1950s, and they eventually won a Nobel Prize for their efforts.

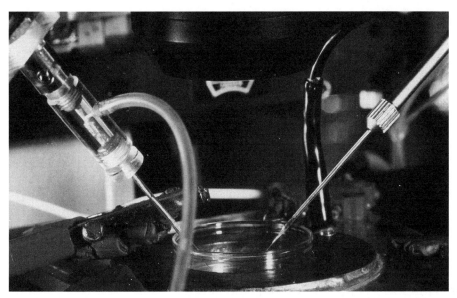

▲ **Figure 9.9** The giant axons of the squid can easily be seen by the naked eye making it possible to manipulate them in experiments.

What happens in an axon?

All cells have a potential difference (electrical voltage) across their surface membrane. Figure 9.10 shows an experimental set-up designed to measure the potential difference across the membrane of an axon.

Topic 9

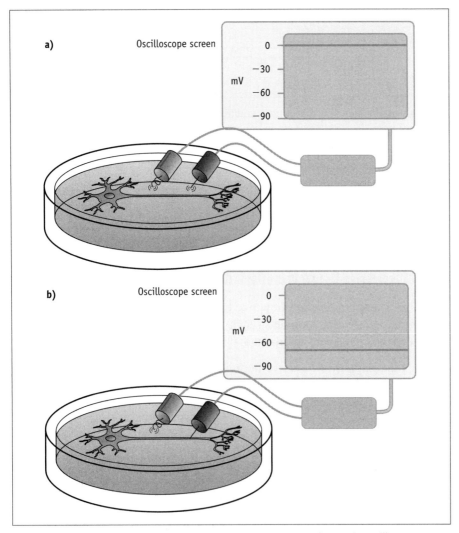

▲ **Figure 9.10** Measuring the potential difference across the axon membrane. The oscilloscope displays the potential difference between the two electrodes.

In Figure 9.10a, with both electrodes in the bathing solution, there is no potential difference. But if one of the electrodes is pushed inside the axon, as in Figure 9.10b, then the oscilloscope shows that there is a potential difference of around −70 millivolts. The membrane is said to be polarised and the inside of the axon is more negative than outside. The value of −70 mV is known as the **resting potential**.

● **Key biological principle:** Why is there a potential difference?

Table 9.1 shows the concentrations of some of the ions found in the solutions inside and outside a squid giant axon. The most obvious feature of this is that the distribution of the ions is far from equal.

This uneven distribution of ions across the membrane is achieved by the action of sodium–potassium pumps in the cell surface membrane of the axon. These carry Na^+ out of the cell and K^+ into the cell. These pumps

Table 9.1 This table shows the concentrations of ions inside and outside a nerve fibre (mmol kg^{-1}). From Hodgkin 1958.

Ion	Extracellular concentration	Intracellular concentration
K$^+$	30	400
Na$^+$	460	50
Cl$^-$	560	100
Organic anions	0	370

act against the concentration gradients of these two ions and are driven by energy supplied by hydrolysis of ATP. The organic anions (e.g. amino acids) are large and stay within the cell, so chloride ions move out of the cell to partly balance the charge across the cell surface membrane.

The membrane is virtually impermeable to sodium ions, but is permeable to potassium ions. As the concentration gradients are established by the sodium–potassium pumps, potassium ions diffuse out of the cell. They pass through potassium channels, making the outside of the cell surface membrane positive and the inside negative. The resting potential is the result of the uneven distribution of K$^+$ ions across the membrane.

So why is the axon resting potential -70 mV? To understand this, we need to appreciate that there are two forces involved in the movement of these ions. First there are the concentration gradients generated by the Na$^+$/K$^+$ pump. There is also the effect of the electrical gradient due to the difference in charge on the two sides of the membrane. Potassium ions diffuse out of the cell due to the concentration gradient. The increased negative charge created inside the cell as a consequence attracts potassium ions back across the membrane into the cell. When the potential difference across the membrane is around -70 mV, the number of potassium ions being attracted into the cell by the electrical gradient balances the diffusion of K$^+$ ions out of the cell. There is no net movement of K$^+$, and hence a steady state exists, maintaining the potential difference at -70 mV. An electrochemical equilibrium for potassium ions is in place, and the membrane is polarised (Figure 9.11).

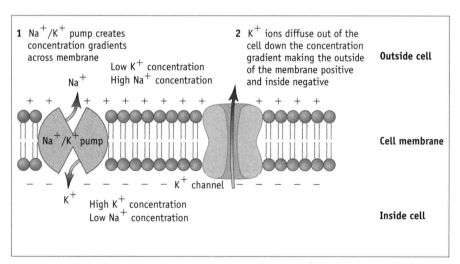

▲ **Figure 9.11** The resting potential. At -70 mV the movement of K$^+$ into the cell due to the electrical potential difference exactly balances the movement out, due to the diffusion gradient. There is also a small amount of Na$^+$ leakage into the cell.

Topic 9

What happens when a nerve is stimulated?

Neurones are electrically excitable cells, which means that the potential difference across their cell surface membrane change when they are conducting an impulse.

Figure 9.12 shows the effect of stimulating the axon by passing a small electric current between two electrodes. If an electrical current above a threshold level is applied to the membrane, it causes a massive change in the potential difference. The potential difference across the membrane is locally reversed, making the inside of the axon positive compared with the outside. This is known as **depolarisation**. The potential difference becomes +40 mV or so for a very brief instant, lasting about 3 milliseconds (ms), before returning to the resting state, as shown by the oscilloscope trace. It is important that the membrane is returned to the resting potential as soon as possible in order that more impulses can be conducted. This return to a resting potential of −70 mV is known as **repolarisation**. The large change in the voltage across the membrane is known as an **action potential** (Figure 9.13).

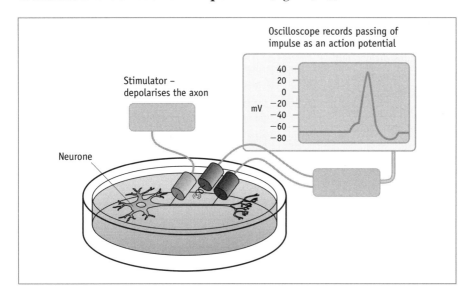

◀ **Figure 9.12** Measuring an action potential.

What causes an action potential?

Once threshold stimulation occurs, an action potential is caused by changes in the permeability of the cell surface membrane to Na^+ and K^+, due to the opening and closing of voltage-dependent Na^+ and K^+ channels (Figure 9.13). At the resting potential, these channels are blocked by gates preventing the flow of ions, as shown in Figure 9.13. Changes in the voltage across the membrane cause the channels to open, and so they are referred to as voltage-dependent gated channels (you get a good score for that in Scrabble).

When a neurone is stimulated, a change in the potential difference across the membrane causes a change in the shape of the Na^+ gate, opening some of the voltage-dependent sodium channels. As the sodium ions flow in, depolarisation increases, triggering *more* gates to open once a certain threshold is reached. This increases depolarisation further. This is an example of **positive feedback**: change encourages further change, and it leads to a rapid opening of *all* of the Na⁺ gates. This means there is no means of controlling the degree of depolarisation of the membrane; action potentials are either there or they are not. This property is often referred to as **all-or-nothing**.

Q9.9 Why is the Na⁺ channel described above referred to as voltage-dependent?

Due to the higher concentration of sodium ions outside of the axon, Na⁺ flows rapidly inwards, causing a build-up of positive charges inside. This reverses the polarity of the membrane. The potential difference across the membrane reaches +40 mV.

After about 0.5 ms, the voltage-dependent Na⁺ channels spontaneously close and Na⁺ permeability of the membrane returns to its usual very low level. Voltage-dependent K⁺ channels open due to the depolarisation of the membrane, and potassium ions move out of the axon, down the electrochemical gradient (they diffuse down the concentration gradient and are also attracted by the negative charge outside the cell surface membrane). As K⁺ ions flow out of the cell, the inside of the cell once again becomes more negative than the outside. This is the falling phase of the oscilloscope trace in Figure 9.13. The membrane is very permeable to potassium, and more ions move out than occurs at resting potential, making the potential difference more negative than the normal resting potential: the membrane is hyperpolarised. The voltage-dependent K⁺ channels close again, and the resting potential is re-established by potassium diffusion into the cell.

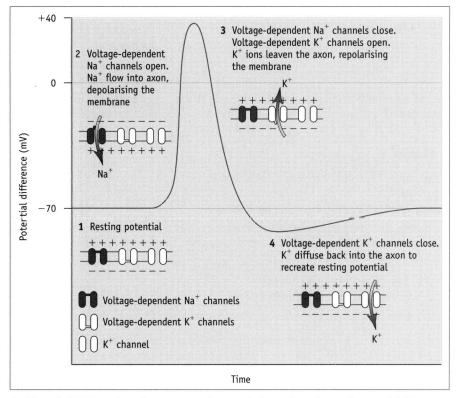

▲ **Figure 9.13** Voltage-dependent gates opening and closing produce changes in potential difference during an action potential.

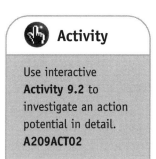

Activity

Use interactive **Activity 9.2** to investigate an action potential in detail.
A209ACT02

Topic 9

After lots (hundreds) of action potentials have occurred in the neurone, the sodium concentration inside the cell rises significantly. The sodium–potassium pumps start to function, restoring the original ion concentration across the cell membrane (Table 9.1). If a cell is not transmitting many action potentials, the pumps will not have to be used very frequently. At rest there is some slow leakage of sodium ions into the axon. These sodium ions are pumped back out of the cell.

Q9.10 Will it be possible for an action potential to be triggered in a dead axon? Give a reason for your answer.

How is the impulse passed along an axon?

When a neurone is stimulated, the action potential generated does not really travel along the axon, but triggers a sequence of action potentials along its length. It is rather like pushing one domino to topple a whole line of standing dominoes. Figure 9.14 illustrates this propagation of the impulse along an axon. As part of the membrane becomes depolarised at the site of an action potential, a local electrical current is created as the electrically charged sodium ions flow between the depolarised part of the membrane and the adjacent resting region. The depolarisation spreads, and the nearby Na^+ gates will respond to this by opening as we described earlier, triggering another action potential. These events are then repeated along the membrane. As a result, a wave of depolarisation will pass along the membrane. This is the nerve impulse.

A new action potential cannot be generated in the same section of membrane for a millisecond or two. This is known as the **refractory period**. It lasts until all the voltage-dependent sodium and potassium channels have returned to their normal resting state (closed) and the resting potential is restored.

Activity

Use **Activity 9.3** to help understand how the nerve impulse is transmitted along the axon. **A209ACT03**

Extension

Read about ion channels and episodic diseases in **Extension 9.1**. **A209EXT01**

Q9.11 How will the refractory period ensure that the action potential will not be propagated back the way it came?

Are impulses different sizes?

The all-or-nothing effect for action potentials means that the size of the stimulus, assuming it is above the threshold, has no effect on the size of the action potential, only the frequency of discharge. A very strong light will produce the same size action potential in a neurone from your eye as does a dim light. Different mechanisms must be used to communicate the intensity of a stimulus; the frequency of impulses is used for this purpose along with the number of neurones in a nerve that are conducting impulses. A high frequency and the firing of many neurones are usually associated with a strong stimulus.

Speed of conduction

The speed of nervous conduction is in part determined by the diameter of the axon. The normal axons of a squid (diameter 1–20 μm) conduct impulses at around 0.5 ms^{-1}, whereas the giant axons (diameter up to 1000 μm) conduct

Grey matter

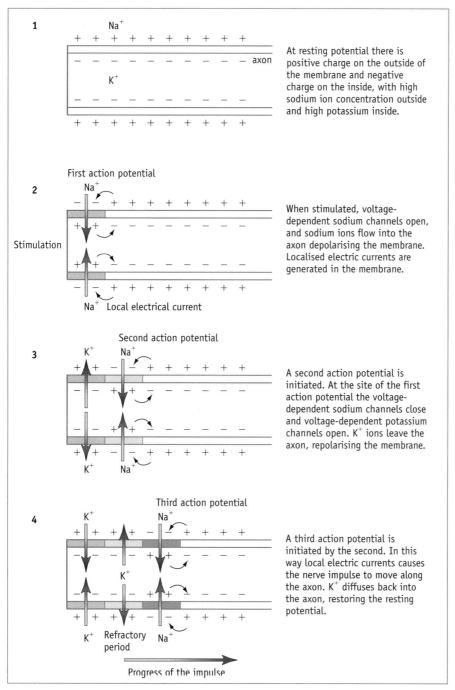

▲ **Figure 9.14** Propagation of an impulse along an axon.

at nearer 100 ms^{-1}. However, the nerve axons of mammals are much finer than the squid giant axons, about 1–20 μm in diameter, but impulses travel along them at up to 120 ms^{-1}. This apparent anomaly can be explained by the presence of the myelin sheath around mammalian nerve axons.

The myelin sheath acts as an electrical insulator along most of the axon, preventing any flow of ions across the membrane. Gaps occur in the myelin sheath at regular intervals (gaps are known as **nodes of Ranvier**), and these are the only places where depolarisation can occur. As ions flow across the membrane at one node during depolarisation, a circuit is set up which reduces the potential difference of the membrane at the next node, triggering an action potential. In this way, the impulse effectively jumps from one node to the next. This is much faster than a wave of depolarisation along the whole membrane. This 'jumping' conduction, illustrated in Figure 9.15, is called saltatory conduction.

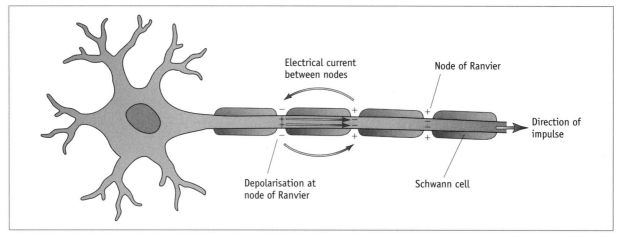

▲ **Figure 9.15** An impulse can move very quickly along the axon by jumping between the nodes of Ranvier.

How does a nervous impulse pass between cells?

Where two neurones meet is known as a **synapse**. The cells do not actually touch – there is a small gap, the **synaptic cleft**. So how does the nerve impulse, on which the function of the nervous system depends, get across this gap?

Synapse structure

A nerve cell may have very large numbers of synapses with other cells, possibly as many as 10 000 for some brain cells. This is important in enabling the distribution and processing of information.

Figure 9.16 shows the structure of a typical synapse. Notice the synaptic cleft that separates the **presynaptic membrane** of the stimulating neurone from the **postsynaptic membrane** of the other cell. The gap is about 20–50 nm and a nerve impulse cannot jump across it. In the cytoplasm at the end of the presynaptic neurone there are numerous **synaptic vesicles** containing a chemical called a **neurotransmitter**.

How does the synapse transmit an impulse?

The arrival of an action potential at the presynaptic membrane causes the release of the neurotransmitter into the synaptic cleft. The neurotransmitter diffuses across the gap, resulting in events that cause the depolarisation of the postsynaptic membrane, and hence the propagation of the impulse along the next cell.

Many neurotransmitters have been discovered, with 50 identified in the human central nervous system. Acetylcholine, the first to be discovered, will be used here to describe the working of a synapse. Others will be considered later in the topic.

There are essentially four stages leading to the nerve impulse passing along the postsynaptic neurone:

- production and storage of neurotransmitter
- neurotransmitter release
- stimulation of the postsynaptic membrane
- inactivation of the neurotransmitter.

These stages are illustrated in Figure 9.16.

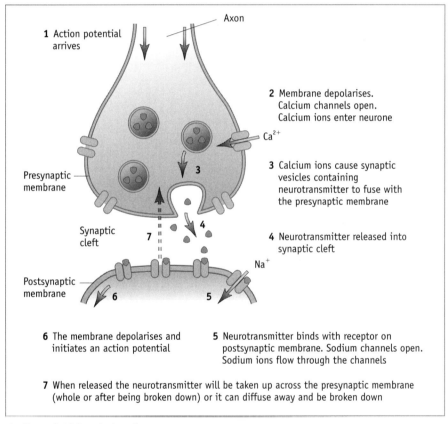

▲ **Figure 9.16** Functioning of a synapse.

- Production and storage of neurotransmitter

    The presynaptic cell expends a considerable amount of energy to produce neurotransmitter and to package it into vesicles ready for transport out of the cell.

- Neurotransmitter release

    When the presynaptic membrane is depolarised by an action potential, channels in the membrane open and increase the permeability of the membrane to calcium ions (Ca^{2+}). These calcium ions are in greater concentration outside the cell, so they diffuse across the membrane and into the cytoplasm.

The increased Ca^{2+} concentration causes synaptic vesicles to fuse with the presynaptic membrane and release their contents into the synaptic cleft by exocytosis.

- Stimulation of the postsynaptic membrane

The neurotransmitter takes about 0.5 ms to diffuse across the synaptic cleft and reach the postsynaptic membrane. Embedded in the postsynaptic membrane are specific receptor proteins that have a binding site with a complementary shape to part of the acetylcholine molecule. The acetylcholine molecule binds to the receptor, changing the shape of the protein, opening cation channels and making the membrane permeable to sodium ions. The flow of sodium ions across the postsynaptic membrane causes depolarisation, and if there is sufficient depolarisation, an action potential will be produced and propagated along the postsynaptic neurone.

The extent of the depolarisation will depend on the amount of acetylcholine reaching the postsynaptic membrane. This will depend in part on the frequency of impulses reaching the presynaptic membrane. A single impulse will not usually be enough. The number of functioning receptors in the postsynaptic membrane will also influence the degree of depolarisation.

- Inactivation of the neurotransmitter

Some neurotransmitters are actively taken up by the presynaptic membrane as soon as they have been released, and the molecules are used again. In others, the neurotransmitter rapidly diffuses away from the synaptic cleft or is taken up by other cells of the nervous system. In the case of acetylcholine, a specific enzyme, **acetylcholinesterase**, at the postsynaptic membrane breaks down the acetylcholine so that it can no longer bind to receptors. Some of the breakdown products are then reabsorbed by the presynaptic membrane and reused.

> 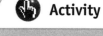 **Activity**
>
> You can investigate the synapse in more detail using the animation in **Activity 9.4. A209ACT04**

What is the role of synapses in nerve pathways?

Synapses slow down nerve impulses, so for rapid responses reflex pathways have as few as possible. However, there are benefits from their presence, including:

- control of nerve pathways allowing flexibility of response
- integration of information from different neurones allowing a co-ordinated response.

The postsynaptic cell is likely to be receiving input from many synapses at the same time (Figure 9.17). It is the overall effect of all of these synapses that will determine whether the postsynaptic cell generates an action potential. Two main factors affect the likelihood that the postsynaptic membrane will depolarise: the type of synapse, and number of impulses received.

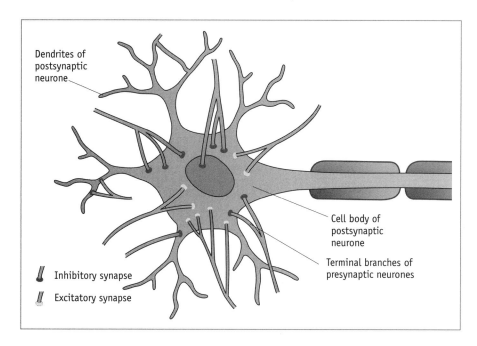

Figure 9.17 The postsynaptic neurone receives input from many excitatory and inhibitory synapses at the same time.

Some synapses help stimulate an action potential, whereas others are inhibitory. They make it *less* likely that the postsynaptic membrane will depolarise.

Excitatory synapses make the postsynaptic membrane more permeable to sodium ions. A single excitatory synapse typically does not depolarise the membrane enough to produce an action potential, but several impulses added together produce sufficient depolarisation via the release of neurotransmitter to produce an action potential in the postsynaptic cell. The fact that each impulse adds to the effect of the others is known as **summation**. If the impulses are from different synapses, usually from different neurones, this is known as **spatial summation**. In this way, the fact that a number of different sensory cells have been stimulated can be taken into account in the control of the response.

When several impulses arrive at a synapse having travelled along a single neurone, each will cause the release of neurotransmitter. Their combined effect may be to generate an action potential in the postsynaptic membrane. This is known as **temporal summation.** Generally, the more intense the stimulus detected by a sensory cell, the more impulses it will generate. In this way more intense stimuli are more likely to cause a response. One small insect landing on your arm may not be noticed, but a larger one crawling along your arm would stimulate lots of receptors over a longer period of time, causing you to respond.

Inhibitory synapses make it less likely that an action potential will result in the postsynaptic cell. The neurotransmitter from these synapses opens channels for chloride ions and potassium ions in the postsynaptic membrane, and these will then follow their diffusion gradients. Chloride ions will move into the cell carrying negative charge and potassium ions will move out carrying positive charge. The result will be a *greater* potential difference across the membrane as the inside will become more negative than usual (about −90 mV), so-called hyperpolarisation. This makes depolarisation less likely.

A postsynaptic cell can have many inhibitory and excitatory synapses, and so whether or not an action potential results depends upon the balance of excitatory and inhibitory synapses acting at any given time.

9.2 Reception of stimuli – how does light trigger nerve impulses?

When reflected light entered Kenge's eye, how was the light converted into electrical impulses which could be passed along the optic nerve to the brain?

Stimuli (any changes that occur in an animal's environment) are detected by receptor cells that send electrical impulses to the central nervous system. Many receptors are spread throughout the body, but some types of receptor cells are grouped together into **sense organs**. Sense organs such as eyes help to protect the receptor cells and improve their efficiency; structures within the sense organ ensure that the receptor cells are able to receive the appropriate stimulus. The receptor cells that detect light are found in the eye. The lens and cornea refract (bend) the light so that it is focused on the retina where the photoreceptor cells are located.

 Nice to know: Different types of receptors

Receptors allow us to perceive and respond to a wide variety of stimuli. The receptors can either be cells that synapse with a sensory neurone, or can be part of a specialised sensory neurone, like the temperature receptors in the skin (see Figure 7.36, Student book 3 page 108). Four of the main types of receptor are shown in the table below.

Type of receptor	Stimulated by	Example of role in body
Mechanoreceptors	Forces that stretch, compress or move the sensor	Balance, touch and hearing
Chemoreceptors	Chemicals	Taste, smell and regulation of chemical concentrations in the blood
Thermoreceptors	Temperature	Thermoregulation and awareness of changes in the surrounding temperature
Photoreceptors	Light	Sight

All of the receptors, except the photoreceptors, work in a similar manner. At rest, the cell surface membrane has a negative resting potential. Stimulation of the receptor causes depolarisation of the cell. The stronger the stimulus the greater the depolarisation. When depolarisation exceeds the threshold level, it triggers an action potential. This is either relayed across the synapse using neurotransmitters or passed directly down the axon of the sensory nerve.

Before addressing in detail the question 'How does light trigger nerve impulses?' remember what you've learnt before about the way that the parts of the eye work together, using Figure 9.18 and the revision quiz in Activity 9.5.

> **Activity**
>
> Remind yourself about the structure and function of the different parts of the eye in **Activity 9.5**.
> **A209ACT05**

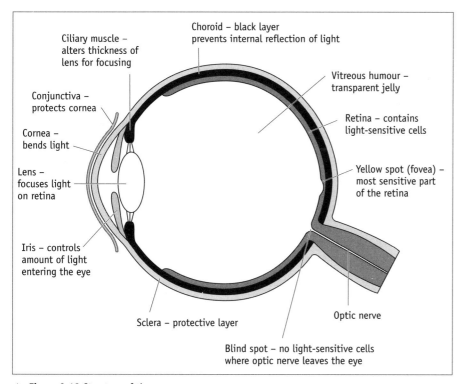

▲ **Figure 9.18** Structure of the eye.

How does light stimulate photoreceptor cells?

The human retina contains two types of photoreceptor cells sensitive to light: **rods** and **cones** (see Figure 9.19). Cones allow colour vision in bright light; rods only give black and white vision, but, unlike cones, work in dim light conditions. In the centre of the retina, in an area about the size of this 'o', there are only cones. This area is responsible for the most detailed pinpointing of the source and nature of a viewed object. Over the remainder of the retina, the rods outnumber the cones in a ratio of about 20 to 1.

In Figure 9.19 notice the arrangement of the three layers of cells that make up the retina; light hitting the retina has to pass through the layers of neurones *before* reaching the rods and cones. The rods and cones synapse with bipolar neurone cells, which in turn synapse with ganglion neurones, whose axons together make up the optic nerve.

Q9.12 Can you explain why some people describe the retina as *functionally inside out*?

In both rods and cones, a photochemical pigment absorbs the light resulting in a chemical change. In rods this is a reddish pigment called **rhodopsin**. In Figure 9.20 you can see that the rod cell contains many layers of flattened vesicles. The rhodopsin molecules are located in the membranes of these vesicles.

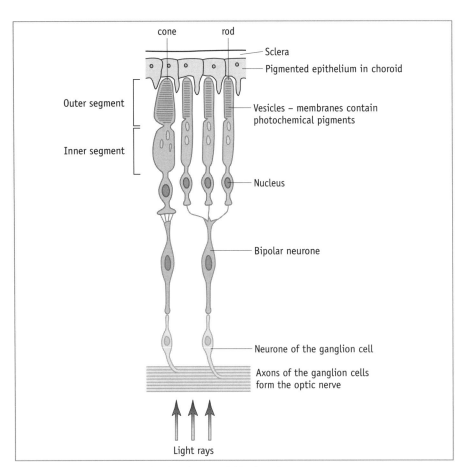

▲ **Figure 9.19** The structure of rods and cones within the retina.

▲ **Figure 9.20** An electron micrograph of rods shows the outer segments that contain the photochemical pigment.

Look back at the diagram of the rod cell and notice that it has an inner and outer segment. In the dark, sodium flows into the outer segment through **non-specific cation channels**. The sodium ions move down the concentration gradient into the inner segment where pumps continuously transport them back out of the cell. The influx of Na^+ produces a slight depolarisation of the cell. The potential difference across the membrane is about −40 mV compared to the −70 mV resting potential of a cell. This slight depolarisation triggers the continual release of a neurotransmitter, thought to be glutamate, from the rod cells. In the dark, rods release this neurotransmitter continuously. The neurotransmitter binds to the bipolar cell, stopping it from depolarising.

When light falls on the rhodopsin molecule, it breaks down into retinal and opsin. The opsin activates a series of membrane-bound reactions which ends with hydrolysis of a molecule attached to the cation channel in the outer segment. The hydrolysis results in the closing of the cation channels. Na^+ influx into the rod decreases, while the inner segment continues to pump Na^+ out. This makes the inside of the cell more negative. It becomes **hyperpolarised,** and the release of the neurotransmitter glutamate stops. As a result, the bipolar cell with which the rod synapses becomes depolarised as cation channels in the membrane open – the neurotransmitter normally prevents them opening. The neurones that make up the optic nerve are also depolarised and respond by producing an action potential. Figure 9.21 summarises the processes that occur in the light and dark.

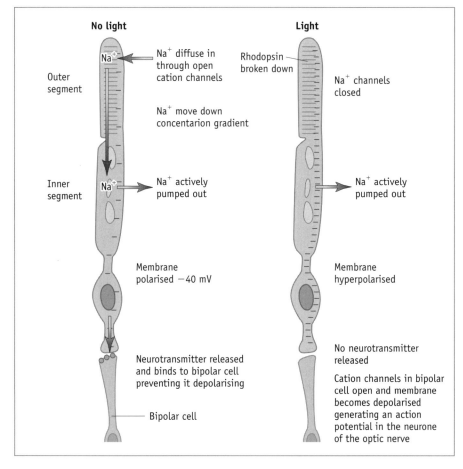

◀ **Figure 9.21** Rod cell in light and dark.

Q9.13 a) By what form of transport will sodium ions be i) pumped out of the rod cell and ii) flow back into the cell?
b) What do you think a 'non-specific cation channel' is?
c) Why does the rod cell membrane become hyperpolarized in the light?

Once the rhodopsin has been broken down, it is essential that it be rapidly converted back to its original form so that repeated stimuli can be perceived. Each individual rhodopsin molecule takes a few minutes to do this. The higher the light intensity, the more rhodopsin molecules are broken down and the longer it can take for all the rhodopsin to reform, up to a maximum of 50 minutes. This reforming of rhodopsin is called **dark adaptation**.

 Activity

Try **Activity 9.6** to experience the effects of dark adaptation, and think through how the rod cells work. **A209ACT06**

Processing of the signals (nerve impulses) from the photoreceptors starts within the retina. The bipolar cells receive signals from the photoreceptors, but may also receive signals from the adjacent bipolar cells. If photoreceptors are stimulated they inhibit their neighbours. Higher intensity light has a greater inhibiting effect. This is called **lateral inhibition**.

Topic 9

> ### Nice to know: Why you should eat your carrots
>
> Have you ever been told to finish your carrots so that you will be able to see in the dark? And did you believe it? Like many old wives' tales there is a grain of truth in this one.
>
> Poor night vision, sometimes called night blindness, has been known for many years to be one of the symptoms of the disease caused by a shortage of vitamin A in the diet.
>
> **Q9.14** What do you think you would need to eat in order to increase the amount of vitamin A (carotene) in your diet?
>
> Vitamin A is closely related to a substance called retinal, which is part of the rhodopsin found in the rods. A shortage of vitamin A leads to a lack of retinal and thus rhodopsin, which means poor vision in low light conditions.

> **Extension**
>
> Read **Extension 9.2** to find out how the cones function to allow colour vision.
> **A209EXT02**

Information from the retina does not inform the brain exactly what the light intensity is at every point. Instead, it highlights where there are changes in the intensity. So the perceived brightness of two adjoining areas is not a reflection of the actual light intensity stimulating the cells in that part of the retina, but an impression of the relative light intensities. This visual contrast can be demonstrated using a simple set of shaded squares, as shown in Figure 9.22.

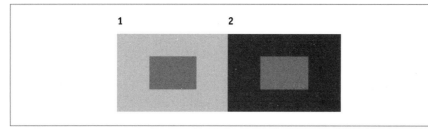

▲ **Figure 9.22** Both central rectangles are the same shade and reflect the same amount of light onto the retina, but due to lateral inhibition they are seen as different shades. The photoreceptors stimulated by light reflected from the area surrounding the central rectangle in 1 receive a greater light intensity. Therefore they inhibit the photoreceptors detecting the central rectangle. This makes the central rectangle appear dimmer. In rectangle 2 the central rectangle reflects more light than the surrounding area; therefore the photoreceptors detecting the central rectangle inhibit the photoreceptors detecting the outer region. The central rectangle appears lighter.

From the eye to the brain

The axons of the ganglion cells that make up the optic nerve pass out of the eye and extend to several areas of the brain including a part of the thalamus as shown in Figure 9.23. The impulses are then sent along further neurones to the primary visual cortex where further processing of the information occurs.

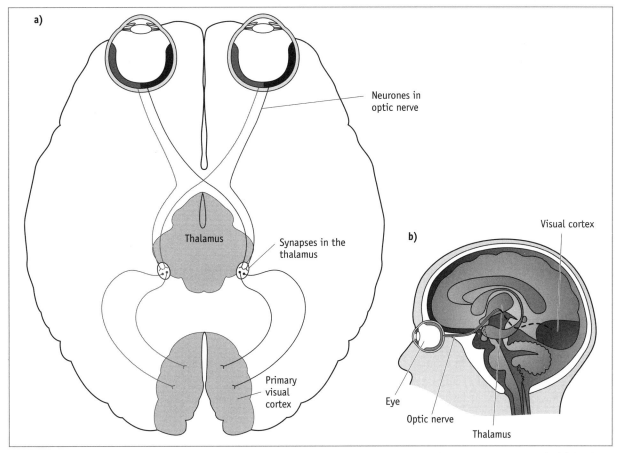

▲ **Figure 9.23** The visual pathway. The right side of the brain interprets input from the right side of both eyes' retinas, i.e. the *left* hand side of the field of view; and the left side of the brain from the left side of both retinas.

Before reaching the thalamus, some of the neurones in each optic nerve branch off to the midbrain, where they connect to motor neurones involved in controlling the pupil reflex and movement of the eye. Audio signals also arrive at the midbrain so we can quickly turn our eyes in the direction of a visual or auditory stimulus.

To discover where the visual cortex is found in the brain and how it develops, read Sections 9.3 and 9.4 before considering what is happening in the visual cortex in Section 9.5.

9.3 Regions of the brain

The brain acts as the main coordinating centre for nervous activity, receiving information from sense organs, interpreting it, and then transmitting information to effectors. Different regions within the brain are involved in helping us respond to our external environment and regulating our internal environment (see Figures 9.24 and 9.25).

Looking at the brain from the top down (Figure 9.24), you see the outer layer, the **cortex**, grey and highly convoluted – the grey matter – composed mainly of nerve cell bodies, synapses and dendrites.

The cortex, accounting for about two-thirds of the brain's mass, is the largest region of the brain. It is positioned over and around most other brain regions, and is divided into left and right **cerebral hemispheres**. Each hemisphere is composed of four regions called lobes: the **frontal**, **parietal**, **occipital** and **temporal** lobes (Figure 9.24a). Each lobe interprets and manages discrete sensory inputs. The two cerebral hemispheres are connected by a sweeping band of white matter (nerve axons) called the corpus callosum.

Activity

Try **Activity 9.7** to gain an impression of the size of the cortex. **A209ACT07**

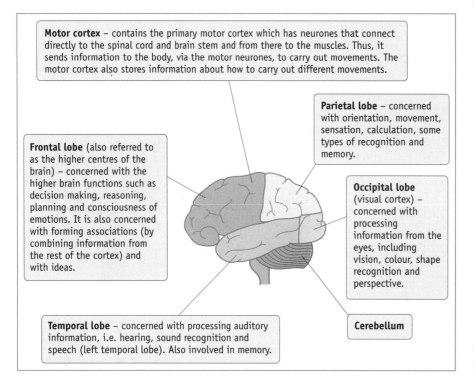

◀ **Figure 9.24** a) The regions of the cerebral hemispheres and their functions.

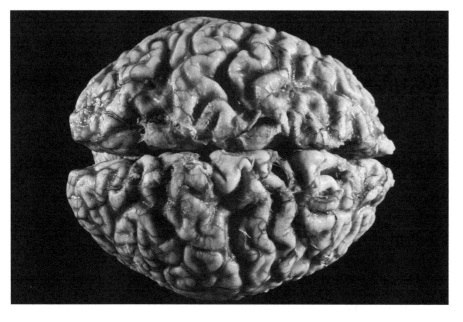

◀ **Figure 9.24** b) The two cerebral hemispheres can easily be seen from above.

Q9.15 A blow to the back of your head may result in you seeing stars. Suggest why.

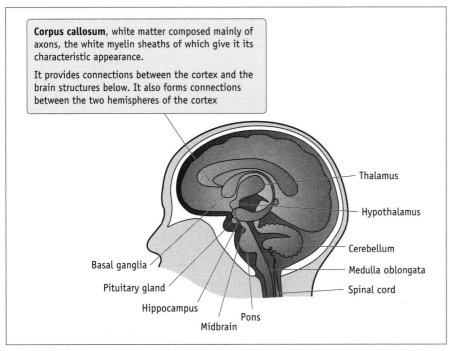

▲ **Figure 9.25** Structures beneath the corpus callosum.

The structures lying directly below the corpus callosum (the white matter) include, among others, the thalamus, the hypothalamus and the hippocampus.

- The **thalamus** is responsible for routeing all the incoming sensory information to the correct part of the brain, via the axons of the white matter.

- The **hypothalamus** lies below the thalamus and contains the thermoregulatory centre. This monitors core body temperature and skin temperature, and initiates corrective action to restore the body to its optimum temperature (see Topics 7 and 8). Also located in the hypothalamus are other centres that control sleep, thirst and hunger. The hypothalamus also acts as an endocrine gland, secreting hormones such as antidiuretic hormone (controls water reabsorption in the kidneys and hence controls blood concentration). The hypothalamus connects directly to the pituitary gland which, in turn, secretes other hormones. See the Key biological principle box on hormonal coordination (page 79).

- The **hippocampus** is involved in laying down long-term memory.

The **basal ganglia** are a collection of neurones that lie deep within each hemisphere and are responsible for selecting and initiating stored programmes for movement.

The cerebellum and brain stem

The brain stem is, in evolutionary terms, the oldest part of the brain and is sometimes referred to as the reptilian brain. It lies at the top of the spinal column. The brain stem extends from the **midbrain** to the **medulla** (Figure 9.25).

Topic 9

Study Table 9.2 which provides information about the role of each region in the brain stem and the **cerebellum**, and then try answering the questions which follow.

▼ **Table 9.2** The cerebellum and brain stem functions.

Cerebellum	Responsible for balance. Coordinates movement as it is being carried out, receiving information from the primary motor cortex, muscles and joints. Constantly checks whether the motor programme being used is the correct one, for example by referring to the incoming information about posture, changing external circumstances.
Midbrain	Relays auditory information to the temporal lobe, and visual information to the occipital lobe.
Medulla	Responsible for regulating those body processes that we do not consciously have to control, such as heart rate, breathing and blood pressure.

Q9.16 Imagine that you are whizzing downhill on a bike and come across an unexpected sharp bend in the road. You need to apply the brakes or turn the handlebars to stop yourself falling off. Which regions of the brain, including those in the brain stem, are involved in your subsequent action?

Q9.17 Various diseases or conditions can give us insights into the functioning of the different areas of the brain. Research has shown that in Parkinson's disease neurones in a particular area of the brain have died. Parkinson's disease results in an inability to select and make appropriate movements. Suggest which area of the brain is damaged.

How do neuroscientists know the function of the different regions of the brain?

How do neuroscientists identify the different regions of the brain and know what each does? Until relatively recently, neuroscientists were only able to study the brain by looking at pathological specimens, by examining the effect of damage to particular areas of the brain, through studies using animal models (see page 85) and, to some degree, by studying human patients during surgery

Studies of individuals with damaged brain regions

By studying the consequences of accidental brain damage it is possible to determine the functions of certain regions of the brain. Researchers have also studied the consequences of injuring or destroying neurones to produce lesions (areas of tissue destruction), and the consequences of the removal of brain tissue in non-human animal 'models'.

Key biological principle: Comparing nervous and hormonal coordination

The nervous system is not the only means by which the activities of the body can be coordinated. Chemicals called **hormones** act as a means of chemical communication with target cells. They are secreted by **endocrine glands** (Figure 9.26) into the bloodstream and transported throughout the body. Each hormone affects only specific target cells, modifying their activity. Some hormones bind to receptors on the cell surface, producing a second messenger that can activate enzymes within the cell; others act on the cell by indirect or direct signalling to control transcription of enzymes. Many hormones are produced steadily over long periods of time to control long-term changes in the body, such as growth and sexual development. See, for instance, testosterone in Topic 8. Adrenaline, also encountered in Topic 8, is more short-term in its action, but is slow in producing a response compared with the action of the nervous system.

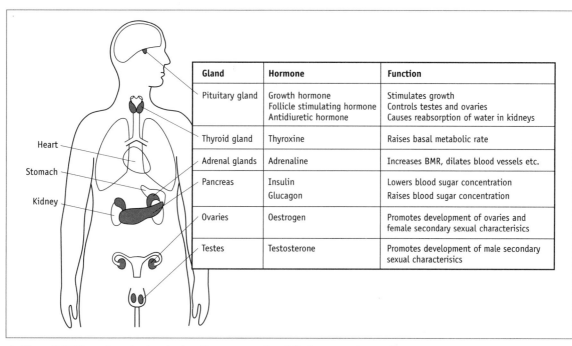

Gland	Hormone	Function
Pituitary gland	Growth hormone Follicle stimulating hormone Antidiuretic hormone	Stimulates growth Controls testes and ovaries Causes reabsorption of water in kidneys
Thyroid gland	Thyroxine	Raises basal metabolic rate
Adrenal glands	Adrenaline	Increases BMR, dilates blood vessels etc.
Pancreas	Insulin Glucagon	Lowers blood sugar concentration Raises blood sugar concentration
Ovaries	Oestrogen	Promotes development of ovaries and female secondary sexual characterisics
Testes	Testosterone	Promotes development of male secondary sexual characterisics

▲ **Figure 9.26** The main endocrine glands with some examples of the hormones they produce.

The table below contrasts nervous and hormonal control in animals.

Nervous control	Hormonal control
Electrical transmission by nerve impulses and chemical transmission at synapses	Chemical transmission through the blood
Rapid acting	Slower acting
Usually associated with short-term changes, e.g. muscle contraction.	Can control long-term changes, e.g. growth
Action potentials carried by neurones with connections to specific cells	Blood carries the hormone to all cells, but only target cells are able to respond
Response is often very local, such as a specific muscle cell or gland	Response may be widespread, such as in growth and development

Activity

In **Extension 9.3** read about how scientists link sex hormones with brain activities.
A209EXT03

Topic 9

The story of Phineas Gage

Phineas Gage was the foreman of a railway construction company; a hard-working, fit, popular and responsible man. One day in 1848 he was working with dynamite when an explosion propelled a three and a half foot long iron bar through his head (Figure 9.27). Amazingly Gage didn't die and although most of the front part of the left-hand side of his brain was destroyed, he could still walk and talk.

But after the accident Gage's personality changed: he became nasty, foul-mouthed and irresponsible. He was also impatient and obstinate and was unable to complete any plans for future action.

Phineas Gage died twelve years later. Researchers at the Harvard University Medical School have since combined photographs and X-rays of Gage's skull with computer graphics to determine the areas of his brain that would have been damaged by the rod, It is highly probable that the accident severed connections between his midbrain and frontal lobes. Gage's reduced ability to control his emotional behaviour after the accident was related to damage at this site.

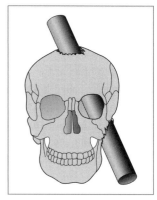

▲ **Figure 9.27** The iron bar travelled behind Gage's left eye and flew out through the top of his skull.

The strange case of Lincoln Holmes

Imagine what it would be like to *never* be able to put a name to a face. That's what it's like for Lincoln Holmes. He finds recognising a face impossible. Thirty years ago a car accident left him with damage to an isolated part of his temporal lobe and he is now 'face-blind'. Even when shown a photograph of himself, he has to be prompted before he realises he is staring at his own image. Lincoln can see facial features, but they all appear as a jumble and he is unable to put all the component parts together. Lincoln's case has revealed that recognition of faces is at least partly carried out by a specific face recognition unit, in the temporal lobe.

Weblink

Listen to Lincoln Holmes on the BBC News website. Damage to an area of his temporal lobe left him 'face-blind'.

Q9.18 Physical damage is one obvious cause of brain damage. Can you think of any others?

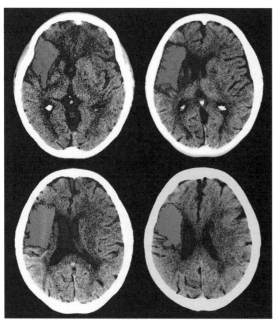

◀ **Figure 9.28** Massive brain lesion in the left hemisphere due to a stroke, affecting the language areas. Surrounding areas are still intact, allowing the patient to sing but not to speak.

Activity

In **Activity 9.8** you can identify regions of the brain by considering symptoms that occur after damage to that area. **A209ACT08**

Brain damage caused by a stroke (Figure 9.28) can cause an impaired ability to speak, trouble understanding speech, and difficulty with reading and writing. In the 19th Century Paul Broca concluded from several post-mortems of patients who could not speak due to strokes that lesions in a small cortical area in the left frontal lobe (subsequently known as Broca's area) were responsible for deficits in language production.

Some patients can recover some abilities after a stroke, showing the potential of neurones to change in structure and function. This is known as **neural plasticity**. The structure of the brain remains flexible even in later life and can respond to changes in the environment. Brain structure and functioning is affected by both nature and nurture.

Individuals with brain damage are still important, providing valuable information in the study of the brain. However, neuroscientists and neurologists now have a wide range of non-invasive imaging techniques for studying the function of the living brain.

Brain imaging

CATs

Computerised Axial Tomography (**CAT** or just **CT**) imaging was developed in the 1970s to overcome the limitations of X-rays. Standard broad-beam X-rays cannot be used for imaging soft tissue, such as the internal structures of the brain, as they are only absorbed by denser materials such as bone.

CAT scans use thousands of narrow beam X-rays rotated around the patient to pass through the tissue from different angles. Each narrow beam is attenuated (reduced in strength) according to the density of the tissue in its path. The X-rays are detected and are used to produce an image of a thin slice of the brain on a computer screen in which the different soft tissues within the brain can be distinguished (Figure 9.28).

CAT scans give only 'frozen moment' pictures. They look at structures in the brain rather than at functions, and are used to detect brain disease and to monitor the tissues of the brain over the course of an illness. However, they have relatively low resolution especially in the brain stem and spinal cord.

Techniques that do not rely on harmful X-rays and can therefore be used more frequently have been developed, including magnetic resonance imaging.

Magnetic resonance imaging

In Topic 1 we discovered how **Magnetic Resonance Imaging** (**MRI**) scans helped to diagnose Mark Tolley's stroke. Here we consider MRI in a little more detail.

MRI uses a magnetic field and radio waves to detect soft tissues. Different tissues respond differently, and so produce contrasting signals and distinct regions on the image. MRI examines tissues in small sections, normally thin 'slices' which when put together by a computer can give three-dimensional images.

Activity

In **Activity 9.9** you use brain images to identify the different regions of the brain and their functions. Visit some of the websites associated with the activity to see lots of CAT and MRI scans.
A209ACT09

Extension

Use **Extension 9.4** to find out in more detail how the brain imaging techniques work. **A209EXT04**

Nowadays MRI is widely used in the diagnosis of tumours, strokes, brain injuries and infections of the brain and spine. MRI can be used to produce finely detailed images of brain structures, as shown in Figure 9.29 with better resolution than CT scans for the brain stem and spinal cord.

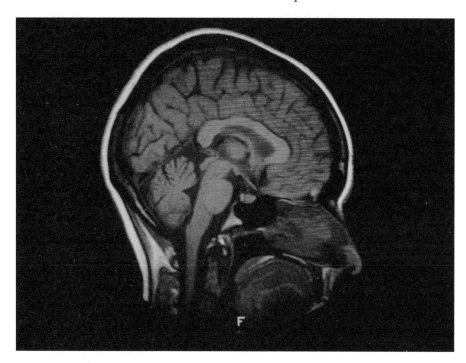

◀ **Figure 9.29** Typical MRI image showing good resolution of soft tissues.

Functional magnetic resonance imaging

Functional Magnetic Resonance Imaging (fMRI) is a particularly useful and exciting tool for the scientist, as it can also provide information about the brain in action. Using this technique it is possible to study human activities, such as memory, emotion, language and consciousness.

fMRI is used to look at the functions of the different areas of the brain by following the uptake of oxygen in active brain areas. This is possible as haemoglobin absorbs the radio wave signal, whereas oxyhaemoglobin does not. Activation of a brain area results in an increased demand for oxygen, and hence an increase in blood flow, and oxyhaemoglobin. The more signal absorbed, the higher the level of activity in a particular area, so different areas of the brain will 'light up' according to when they are active (see some fMRI images on the websites linked to Activity 9.9).

fMRI can produce up to four images per second, so the technique can be used to follow the sequence of events over quite short time periods. In a typical MRI experiment, images are collected continually while the subject alternates between resting and carrying out some task, such as object recognition, listening, or memorising number sequences.

Q9.19 A neuroscientist conducts an fMRI experiment to investigate brain activity when subjects perform voluntary actions like pressing a lever. Look back at the section on regions of the brain and decide which part of the brain would be expected to be active when the lever is being pressed.

Q9.20 A study of a group of London's taxi drivers showed that the right hippocampus is involved in recalling a well-developed mental map of London. A second study examined brain scans of 16 London taxi drivers and found that the only areas of their brains that were different from 50 control subjects were the left and right hippocampus. A particular region, the posterior hippocampus, was found to be significantly larger in the taxi drivers, whilst the front of the hippocampus was smaller than in the control subjects.

a) What imaging method could have been used in i) the first and ii) the second investigation?

b) What do you think the scientists were able to infer from their results?

Nice to know: Another imaging technique

Positron Emission Tomography

Positron Emission Tomography (PET) uses a radio-labelled substance, usually water or oxygen, to examine both structures and functions of the brain. The patient is first injected with the radio-labelled substance which binds to chemicals or receptors in the brain. As the radio-labelled substance decays, it emits positrons (a type of elementary particle). When an area of the brain is active, the neurones in that area have an increased energy use. More oxygen and glucose is required, so there is an increased blood flow in that area. An increase in blood flow will show up on a PET image, as more radio-labelled atoms will be present in that area. In the tissue when a positron collides with an electron, two gamma rays are emitted and these can be picked up by detectors, converted into a signal, and then displayed as an image by the computer (Figure 9.30). Bright spots on the image indicate high levels of neurone activity, dark spots low levels. By running a video sequence of PET images, one after another, changes in activity in various areas of the brain can be followed. However, PET scans can only be done once or twice a year for safety reasons. They are also very expensive so are not used for routine screening.

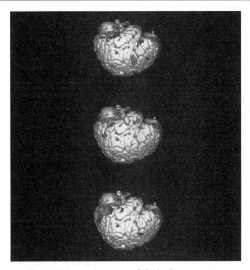

▲ **Figure 9.30** Active areas of the brain appear as coloured areas on a PET scan. These images show the areas of the brain active while processing images of objects. The occipital lobe is active when the objects are seen (top), the temporal lobe lights up as the object is recognised (middle) and the speech and motor area of the frontal cortex are active when the object is named (bottom).

9.4 Visual development

How does the visual cortex develop? What about the connections from the eye to the visual cortex? Was Kenge's visual development in some way faulty, causing him to see the buffalo as insects?

The human nervous system begins to develop soon after conception. By the 21st day, a neural tube has formed. The front part of the neural tubes goes on to develop into the forebrain, midbrain and hindbrain, while the rest of the neural tube develops into the spinal cord. The rate of brain growth during development is truly astonishing. At times, 250 000 neurons are added every minute!

A baby arrives in the world with about 100 billion neurones. There is no huge post-natal increase in number of brain cells after birth though there is a large post-natal increase in brain size. This is caused by several factors, principally elongation of axons, myelination and the development of synapses. By six months after birth the brain will have grown to half its adult size. By the age of two years, the brain is about 80% of the adult size.

Once neurones have stopped dividing, the immature neurons migrate to their final position and start to 'wire themselves'. Neurones must make the correct connections in order for a function such as vision to work properly. Axons grow and synapse with the cell bodies of other neurones.

Axon growth

Axons of the neurones from the retina grow to the thalamus where they form synapses with cells in a very ordered arrangement. Axons from the thalamus grow towards the visual cortex in the occipital lobe. Throughout the nervous system many more neurones are produced than are required, and numerous neurones and axons are pruned back. In the retina up to 80% of the original neurones die.

Staining techniques and studies using electrical stimulation show that the visual cortex is made of columns of cells. Axons from the thalamus synapse within these columns of cells. Adjacent columns of cells receive stimulations from the left and right eye for the same area of the retina with the pattern repeated across the visual cortex (see Figure 9.31). In this way a map of the retina is created within the visual cortex.

It used to be thought that these columns were formed during a critical period for visual development after birth, the result of nurture rather than nature. Periods of time during post-natal development have been identified when the nervous system must obtain specific experiences to develop properly. These are known as **critical periods**, **critical windows** or sensitive periods.

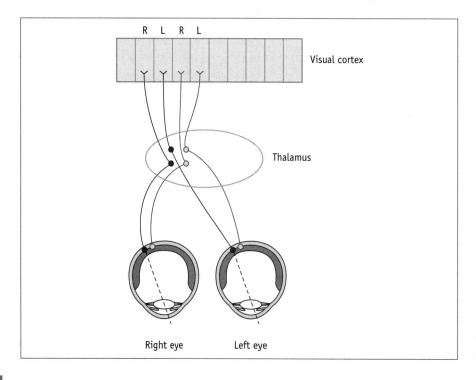

◀ **Figure 9.31** The relationship between the cells in each retina and the cells in the visual cortex.

Evidence for a critical period in visual development

The way in which the environment affects the wiring of the nervous system has been extensively studied using the visual system. The evidence for a critical period in visual development comes from several sources, including medical observations and results of experiments using animals.

Medical observations

One well-known case is that of a young Italian boy who, as a baby, had a minor eye infection. As part of his treatment, his eyes were bandaged for two weeks. When the bandage was removed he was left with impaired vision.

Studies of people born with cataracts have also contributed to our understanding of critical periods in development. You may know that a cataract is the clouding of the lens of the eye (Figure 9.32); this affects the amount of light getting to the retina. If cataracts are not removed before the child is ten years old, they can result in permanent impairment of the person's ability to perceive shape or form, including difficulties in face recognition. By contrast, elderly people who develop cataracts later in life and then have them for several years report normal vision after their removal. This suggests that there is a specific time in development when it is crucial for light stimuli to enter the eye. Cataract removal in children is now carried out at a much earlier stage of development.

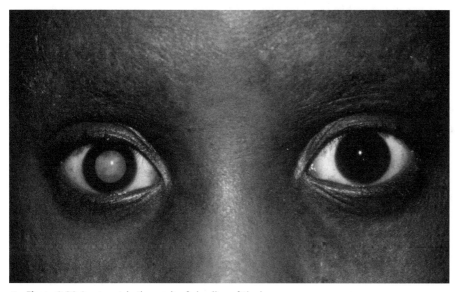

▲ **Figure 9.32** A cataract is the result of clouding of the lens.

Research using animal models

Much of our knowledge and understanding of the visual system, and evidence for critical periods, comes from studies using animals. Animals are used extensively in the study of biological processes including brain development and function. Because most research is conducted on just a few types of animals, a wealth of information is available about them. They are known as animal models.

Topic 9

Most animal models are easy to obtain, easy to breed, have short life-cycles, and small adult size. In Topic 3 we saw fruit flies, *Drosophila*, being used for studying the links between genes and development. Nematode worms, chickens, mice, frogs and zebrafish are also used in this area of research. Mice are used extensively in the study of cancer and disease. In the study of visual development kittens and monkeys have been used, because of their similarity to humans.

Research using animals raises ethical issues – see pages 104–6.

Studies of newborn animals

In one study, one group of newborn monkeys was raised in the dark for the first three to six months of their lives, and another was exposed to light but not to patterns. When the monkeys were returned to the normal visual world, researchers found that *both* groups had difficulty with object discrimination and pattern recognition.

Q9.21 What does this suggest is required for visual development in these monkeys?

In a series of studies, Hubel and Wiesel raised monkeys from birth to six months, depriving them of stimulus to light in one eye. This is known as **monocular deprivation**. After six months the eye was exposed to light. On exposure to light it was clear that the monkey was blind in the light-deprived eye. Retinal cells in the deprived eye did respond to light stimuli, but the cells of the visual cortex did not respond to any visual input from the formerly deprived eye. Deprivation for only a single week during a certain period after birth produced the same result. Deprivation in adults had no effect. Interestingly, visual deprivation of both eyes during this critical window has much less effect than when just one eye is deprived.

Q9.22 Hubel and Wiesel tested kittens for the effects of monocular deprivation at different stages of development and for different lengths of time. They found:
- deprivation at under three weeks had no effect
- deprivation after three months had no effect
- deprivation at four weeks had a catastrophic effect – even if the eye was closed for merely a few hours.

Bearing in mind that kittens are born blind, can you explain the results above?

Q9.23 Young doves and chaffinches raised without exposure to adult song will sing the adult song perfectly the first time they try it – the behaviour is innate (inborn). Most other bird species need exposure to the adult song during a critical period in order learn the proper song. Comment on the main influence on the development of the part of the brain associated with bird song: is it nature or nurture?

Working with ferrets, Crowley and Katz showed by injecting labelled tracers that the columns in the visual cortex are formed before the critical period for development of vision (Figure 9.33). Columns are also seen in newborn monkeys, suggesting that their formation is genetically determined and not the result of environmental stimulation.

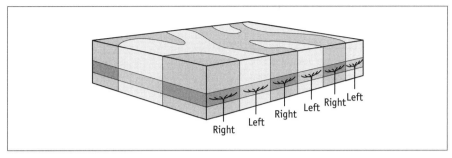

▲ **Figure 9.33** Radioactive label moves from one eye and is concentrated into distinct bands in the visual cortex showing that the columns of cells that receive input from that eye. These banding patterns have been observed in animals that have received no visual stimulation.

What is happening during the critical period for development of vision?

If the columns in the cortex are created before birth what is happening during the critical period which can result in impaired vision if one eye is deprived of light? There must be further visual development. At birth in monkeys there is a great deal of overlap between the territories of different axons (Figure 9.34). In adults, the mass of the brain is greater – there are more dendrites and synapses – but there is less overlap. Are these changes the result of visual experience or genetically determined?

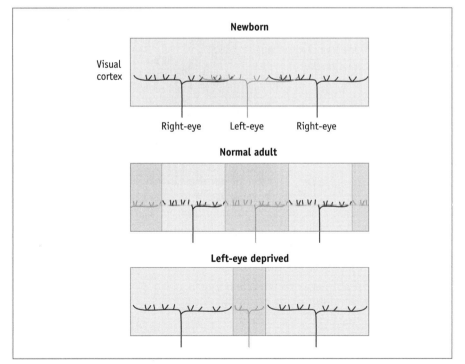

▲ **Figure 9.34** Columns in the visual cortex are present at birth but refinement during the critical period produces the distinctive pattern of columns driven by the left and right eye. Columns that receive input from the light-deprived eye become much narrower.

Topic 9

After light deprivation in one eye, the widths of the columns joining the deprived eye are narrower than for the eye receiving light stimulation (Figure 9.34). Dendrites and synapses from the light-stimulated eye take up more territory in the visual cortex. This suggests that visual stimulation is required for the refinement of the columns, and so for full development of the visual cortex.

Axons compete for target cells in the visual cortex. Every time a neurone fires onto a target cell, the synapses of another neurone sharing the target cell are weakened, and they release less neurotransmitter. If this happens repeatedly, the synapses that are not firing will be cut back. When one eye is deprived of light, the axons in that eye will not be stimulated, so its synapses will be weakened and eventually lost.

 Activity

In **Activity 9.10** you can look at the evidence for a critical period for development of vision. **A209ACT10**

Q9.24 Rewrite these sentences in the correct order, and the pieces of this jigsaw of ideas should fit together.
1 There is a lack of visual stimulation in one eye
2 Inactive synapses are eliminated
3 Axons from the non-deprived eye pass impulses to cells in the visual cortex
4 Synapses made by active axons are strengthened
5 Axons from the visually deprived eye do not pass impulses to cells in the visual cortex.

 Nice to know: Using stem cells to restore vision

Mike May (Figure 9.35), a successful businessman from San Francisco, lost his sight in a chemical explosion when he was three years old. Forty years later, in 2000, he underwent pioneering surgery on his cornea in a bid to restore some vision. Donor stem cells were transplanted into the eye. To be successful the cells needed to multiply and bind to the original cornea cells.

When the bandages came off, to his surprise a limited amount of vision had been restored. However, Mike's biggest problem is understanding what he's seeing. His main method of perception is still through touch. He has to learn how to interpret what he sees. He will never learn to see as clearly as most people.

▲ **Figure 9.35** Mike May who is learning to see again.

9.5 Making sense of what we see

Visual perception is not simply the creation of an image of what is being viewed, but involves prior knowledge and experience as the brain interprets the sensory information received from the retina.

Kenge could see the animals on the plain perfectly well, so his visual development was not a problem. Impulses were successfully sent from the retina to the cortex where cells were stimulated; he then had to make sense of what he saw.

Grey matter

Individual neurones in the visual cortex respond in different ways to the information from the retina, and to different characteristics of the object being viewed. Some neurones, called simple cells, respond to bars of light. Other cells, called complex cells, respond to edges, slits or bars of light that move; others to the angle of the edge; still others to contours, movement or orientation.

Pattern recognition

Look at the list of letters below. You should easily recognise the letters unless you have dyslexia. Some theories of perception suggest that there is a template for patterns which are stored in our long-term memory. The template may not be an exact representation of the letters but a model that has all the key features, so when patterns are viewed their key features are matched with those in the long-term memory, allowing us to recognise and interpret what we see. There is still a lot about dyslexia that we don't know but it clearly has something to do with problems in recognising certain patterns. For example 'b' and 'd' may be confused.

If you look quickly down the list of letters on the right you will probably be able rapidly to pick out the single letter H with its straight vertical and horizontal lines – the other letters all have curved features. You do not have to look at every single letter separately. Cells that respond to the horizontal and straight lines are used; once stimulated they can check that the letter has the features of an H, using a stored template.

When reading we do not have to recognise every single letter. If we did it would take much longer to read this sentence, in the way a small child must sound out each letter in turn. Instead we must be recognising complete words.

Recognition of objects probably works in a similar way. Try Question 9.25 and see if you can recognise the objects. This also shows how the brain makes assumptions about what it expects to see, based on the context and past experience, to make a best estimate at recognition.

Q9.25 Look at the two pictures in Figure 9.36 below, and decide what each is. If you are not sure, have a look over the page for some help.

> **Activity**
>
> In **Activity 9.11** complete a couple of simple experiments that use two areas of the visual cortex.
> **A209ACT11**

CQOOCG
QQDCOQ
GGCOQD
CGQSSU
GCOQUC
ODQUCO
GHOQGD
DGOQDS

> **Activity**
>
> In **Activity 9.12** and **Activity 9.13** you can investigate pattern recognition.
> **A209ACT12**
> **A209ACT13**

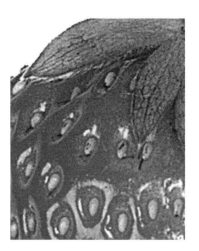 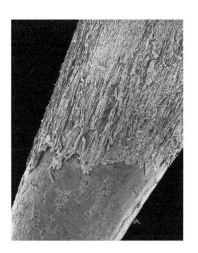

◀ **Figure 9.36** Can you recognise these objects? The answer to Question 9.25 will let you check if you were correct.

Topic 9

▲ Did you guess correctly? See page 89, Question 9.25.

Face recognition

Recent work has shown that we have specialised areas of cells, called recognition units, for identifying complex objects. One of these areas is specifically concerned with face recognition – it is called the face recognition unit. (See the story of Lincoln Holmes on page 80.)

When we first see a new face, certain complex cells in the cortex are stimulated by the lines and contours of the face. These form synapses with other areas of the brain including the language and association areas in the brain. When we see the same face again, we may recognise it, as the necessary neural pathways are already there. In a well-known face, only a few of the visual cues are needed to trigger recognition, as the system is already set up to respond. Look at the fragments of faces in Figure 9.37 and try Question 9.26 to check how many you recognise.

Q9.26 Name the owners of the faces shown below. Explain why you might recognise some but not all of the individuals.

▲ **Figure 9.37** Who are they? Check the answer to Question 9.26 to find out if you are correct.

Grey matter

During development, and indeed throughout our lives, we build up information in the brain that we call upon to interpret sensory information. Information in these libraries is reinforced (and easier to recall) the more times the information is put in. This is why reading your notes and revising thoroughly helps in your exam! You will learn about the molecular basis of this process later in this topic.

Depth perception

For objects less than 30 m away from us, we depend on the presence of cells in the visual cortex that obtain information from both eyes at once. The visual field is seen from two different angles, and cells in the visual cortex let us compare the view from one eye with that from the other. This is called stereoscopic vision.

 Activity

Check if you have stereoscopic vision using **Activity 9.14**.
A209ACT14

Q9.27 Why might a child who had had an eye patch during visual development never develop stereoscopic vision?

The images on our two retinas of objects that are greater than 30 m distant are very similar, so visual cues and past experiences are used when interpreting images. Look at the beach in Figure 9.38. What visual cues might help in depth perception? Lines converge in the distance giving perspective, the impression of distance. The stones further away are smaller and, because from experience we know that the stones along a beach are usually roughly the same size, the small ones in the picture are perceived as being further away. We know the size of buildings, so the ones in the background are not smaller, just further away. Overlaps of objects and changes of colour also help in judging depth. Even if an object becomes smaller on the retina, for example when a car drives away into the distance, we perceive it as moving further away not getting smaller. (But think about how you can tell that a deflating balloon is getting smaller.)

▲ **Figure 9.38** Visual clues in the photograph allow depth perception.

Topic 9

Usually we are unaware of the clues being used. It is only when they are not present, or cause the brain to misinterpret the image, that we become aware of them, such as when viewing optical illusions – see Figure 9.39. The person standing further away may look taller, but the lines that imply depth fool the brain. If you measure the people with the ruler you will see that they are actually the same height.

▲ **Figure 9.39** Are these two people different heights? Check with a ruler to see if you are correct.

Is the perception of the illusion innate or learnt? Nature, nurture or both?

Could it be that Kenge had not developed depth perception over long distances because he had had no experience of seeing long distances on an open plain, having always lived in the enclosed forest? Studies into how people with different cultural experiences and how newborn children respond to optical illusions help answer this question.

Cross-cultural studies

What is culture? It has been defined in a number of different ways. In this course we will view culture as a system of beliefs that are shared among a group of people. It thus shapes experience and behaviour. People from different cultures may not share the same beliefs, and they may show different behaviours.

The Müller-Lyer illusion

The Müller-Lyer illusion may well be familiar to you. Look at Figure 9.40 and decide which of the two vertical lines, A or B, is longer. Both lines are the same length, but most people mistakenly think they differ in length.

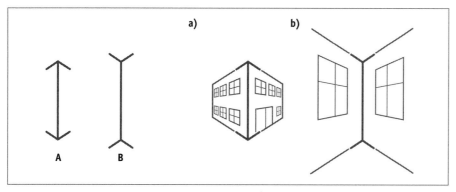

▲ **Figure 9.40** The Müller-Lyer illusion.

According to the **carpentered world hypothesis**, those of us who live in a world dominated by straight lines and right angles tend to interpret images with acute and obtuse angles as right angles. Thus a non-rectangular image, such as a trapezium (a four-sided figure with only two parallel sides), is perceived as rectangular (i.e. having 4 right angles of 90°). This becomes automatic and unconscious from a very early age. For instance, from photographs St Mark's Square in Venice is often perceived to be a rectangle when in fact it is not.

In the Müller-Lyer illusion it has been argued that line A is perceived as the outside corner of a building, but line B as the inside corner of a room (Figure 9.40b). Because we usually look at buildings which are further away than the corner of a room, the room appears larger (look out of the window and, using your fingers, judge the height of a building some distance away; then compare this with the height of the corner of the room you are in). In other words, it is our cultural experience which may account for us perceiving that one line is longer than the other one in this illusion.

 **Activity**

Investigate the Müller-Lyer illusion using the interactive **Activity 9.15**.
A209ACT15

Studies have shown that people who live in a 'circular culture' with few straight lines or corners, such as the Zulu people of Africa who have circular houses and no roads, rarely misjudge the length of the lines in the Müller-Lyer illusion. They have little experience of interpreting acute and obtuse angles on the retina as representations of right angles.

In another study, individuals from a range of different cultures were shown pictures with depth cues such as object size, overlap of objects and linear perspective, similar to those in Figure 9.41.

It was found that all young children had difficulty perceiving the pictures as three-dimensional. They would have said that the man in Figure 9.41 was pointing his finger at the elephant not the antelope – they failed to interpret the depth cues. By the age of 11 years, almost all the European children interpreted the pictures in three dimensions, but some Bantu and Ghanaian children still did not, as did non-literate adults, both Bantu and European. They had had less experience in the interpretation of depth cues in pictures.

What does seem clear is that the depth cues in pictures which most of us take for granted are not innate; they have to be learned. It suggests that visual perception is, in part at least, *learned*.

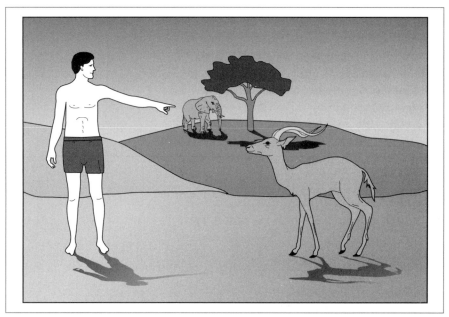

▲ **Figure 9.41** Pictoral depth cues tell us that the man must be pointing at the antelope not the elephant.

Q9.28 **What depth cues are used to decide that the man is pointing at the antelope in Figure 9.41 and not the elephant?**

Some researchers think that lack of susceptibility to the Müller-Lyer illusion is not the result of different experience, but is due to genetic differences in pigmentation between individuals. They suggest that individuals who find it harder to detect contours are less susceptible to the Müller-Lyer illusion. They link poor contour detection to higher retinal pigmentation. In light-coloured people, with low retinal pigmentation, contour detection is good so such people are more easily caught out by the illusion.

Q9.29 **Why might researchers be cautious about the pigment and contour detection evidence supporting the idea that susceptibility to the Müller-Lyer illusion is genetic rather than environmental?**

Q9.30 Suggest reasons why caution should be used when interpreting some of these cross-cultural studies.

> **Activity**
>
> In **Activity 9.16** you can examine some data from cross-cultural studies and think about why Kenge thought the buffalo were insects. **A209ACT16**

Studies with newborn babies

Any inborn capacities exhibited by a newborn baby are used as evidence for the role of genes in determining the hard wiring of the brain before birth. Babies are born with a range of characteristic behaviours which suggest that these are determined by genes; for example crying, walking movements and grasping are all present from birth. Within 24 hours of birth newborn babies can distinguish human faces and voices from other sights and sounds, and prefer them.

Although born short-sighted, babies can see people and items clearly at a distance of about 30 cm. Their preference for stripes and other patterns shows they are imposing order on their perceptions in early infancy. Long before they can crawl, they can tell the difference between a happy face and a sad one. They can imitate people's expressions, and by the time they're old enough to pick up a phone they can mimic what they've seen others doing with it.

The visual cliff

In a classic experiment babies are encouraged to crawl across a table made of glass or perspex below which is a visual cliff. Patterns placed below the glass create the appearance of a steep drop, as shown in Figure 9.42. If the perception of depth is innate, then babies should be aware of the drop even if they have not previously experienced this stimulus themselves.

▲ **Figure 9.42** The visual cliff.

Topic 9

Q9.31 If the baby does have a perception of depth, how will she react when invited to crawl over the 'edge' of the cliff?

Young babies were very reluctant to crawl over the 'cliff' even when their mothers encouraged them to do so.

Q9.32 a) This reaction is assumed to indicate that depth perception is innate. Explain why.
b) Argue the case for this *not* illustrating that depth perception is innate.

This experiment is only possible with babies who have learnt to crawl. It is likely that a six-month-old human may already have learnt depth perception. Therefore the experiment was repeated with animals which can walk as soon as they are born such as chicks, kids (young goats) and lambs. They too refused to cross the cliff.

Q9.33 Explain if this supports or does not support the idea that this type of perception is innate.

Q9.34 Having read section 9.5, can you suggest why Kenge, a forest dweller, was fooled into thinking that the buffalo on the horizon were insects?

9.6 Learning and memory

Kenge laughed when he was told that the animals on the horizon were buffalo so he must have known what a buffalo was. At some point he must have seen a buffalo up close and learnt what it looked like, even though he could not perceive it as one from a distance.

Learning occurs throughout your life, and is any relatively permanent change in behaviour or knowledge that comes from experience. For learning to be effective, you must be able to remember what you have learnt, and studies of learning and memory have always gone hand in hand. Throughout our lives, the memory stores vast amounts of information, from sights and sounds to emotions and skills such as riding a bike or texting. But what is learning and how are memories stored?

Classic studies of learning

When the bell next goes for lunch, think about your mouth. Is it watering? Although you probably had not realised it, you may well have learnt to **associate** the sound of the bell with the arrival of food. This means that the **stimulus** of the bell promotes a **response** to produce saliva. The psychologist, Ivan Pavlov, studied these type of associations with dogs (see Figure 9.43).

▲ **Figure 9.43** By making a sound just before each occasion when the dog received food, the dog learnt to associate the sound with food and would produce saliva on hearing the sound even if no food followed.

> **Activity**
>
> In **Activity 9.17** you can investigate conditioning, including conducting Pavlov's dog experiment yourself, using the animation on the Nobel Prize web site. URL is in the weblinks section for this activity. **A209ACT17**

The dog's behaviour changes relatively permanently. This means that learning has taken place. The sound of the bell is known as a **neutral stimulus** because it would not have promoted salivation before the experiment. The following process has occurred:

- The dog responds to an unconditioned stimulus (receiving food).
- The neutral stimulus (sound) becomes associated with the unconditioned stimulus (receiving food),
- The dog responds to the neutral stimulus whether or not food is present.

This is known as a conditioned response. This type of learning is known as **classical conditioning**.

Topic 9

Q9.35 Name the neutral stimulus, unconditioned stimulus and conditioned response in these examples:
a) in the Pavlov dog experiment
b) in an experiment in which air was puffed into a person's eye at the same time that a sound was heard; eventually the tone alone caused blinking.

Pavlov's dogs showed how conditioning occurs when the neutral stimulus immediately precedes the unconditioned stimulus. If the presentation of food is just as likely with or without the sound preceding it, then no association will develop, or any existing association will be suppressed.

Classical conditioning is not the only form of conditioning. Jayshree was walking along the top of a narrow wall one day, when she lost her footing and fell off. She hurt herself slightly, and her mother shouted 'That'll teach you'. Jayshree can learn from the consequences of her behaviour. This type of learning, illustrated in the experiment shown in Figure 9.44, is an example of **operant conditioning**, where the consequences of an animal's behaviour affect whether that behaviour will occur again. The reward is a **reinforcement**. The reinforcement makes it more likely that the animal will press the lever.

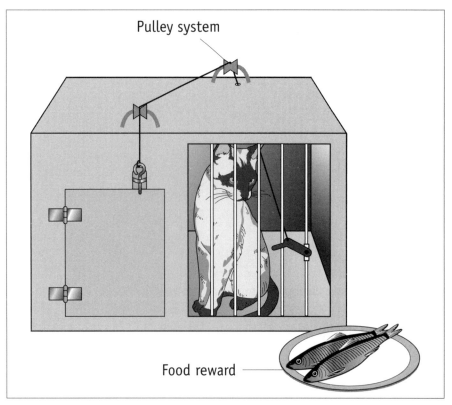

▲ **Figure 9.44** In a classic set of experiments, Edward Thorndike (1898) placed a cat in a box with a plate of fish outside. To get out of the box, the cat had to press a latch, which opened the door. By a process of **trial and error**, the cat learned how to escape from a box. In the early stages, the cat simply made lots of random movements, eventually operating the door latch by accident. Gradually, the cat began to associate the particular movement of operating the latch with food. Having made this association, it could open the door very quickly. It had learned how to escape.

Punishments work in a similar way. If the animal received an electric shock when it pressed the lever, it would soon learn to press less. It would learn that pressing the lever is associated with pain. In the same way Jayshree will learn that falling is painful and she will learn to be more careful.

> ● **Nice to know:** Training animals
>
> By using operant conditioning, you can teach an animal to do remarkably complex behaviours. Imagine you want a rat to stretch up and push a button on the roof of a box with its nose. It is quite unlikely the rat would ever do this by chance. So how do you use reinforcement to achieve your aim? The best method is to break the desired behaviour into stages. For example:
>
> - Reward the (male) rat every time he stretches up, until he is stretching up quite often.
>
> - Stop rewarding for stretching, and reward him every time a part of his body touches the button. He should soon be stretching up and touching the button frequently.
>
> - Stop rewarding for general touching, and only reward him when his nose touches the button. He should soon be touching the button specifically with his nose.
>
> - Stop rewarding for just touching, and only reward him when his nose pushes the button. He should soon frequently be pushing the button with his nose: the desired behaviour.

Not all learning depends on reinforcement. Two case studies are outlined below.

Case study 1: Use your brain!

Wolfgang Köhler conducted experiments on chimpanzees. He placed a banana outside a cage, just out of reach. The chimp saw the banana and sat down. A short time later, it got up, picked up two sticks, joined them together and reached for the banana. Köhler called this type of learning **insightful learning**.

Case study 2: Learn from the actions of others

Learning by **observing** involves watching others and copying them (or deciding not to copy them). If your teacher/lecturer has ever shown you how to do a practical by demonstrating it, you have learnt what to do by watching them. Learning by observing does not appear to involve classical or operant conditioning. Imagine if learning by observing did not take place. Many things would take a lot longer to learn.

Topic 9

There are different types of memory

Take a pencil in your hand and hold it vertically at one end. Move it quickly from left to right, watching as you do so. You should see the image of the pencil trailing behind the pencil. Do it again if you're not sure. This image, which entered your eyes as light, has entered your memory (if only for a split second). It would not appear to be there if you weren't remembering it! This form of memory is called **sensory memory** and lasts for a very short time, only milliseconds.

Having entered your sensory memory, information may be stored in your **short-term memory** (**STM**). There are three main ways information can be stored: semantic (meanings), visual (pictures) and auditory (sounds). For example the word bike could be stored as a picture or visual image of the word, as a sound by saying the word aloud or as a meaning related to the word such as when you ride the bike.

The STM has a limited capacity. Many researches believe it can only hold between five and nine different pieces of unconnected information at once. Information will decay or fade out of STM relatively quickly. Estimates on how long information remains in STM range from 4 seconds to a couple of minutes. Indeed, you can probably not remember the exact wording of the top line of this paragraph, even though you only read it about 10 seconds ago! To keep information in STM (and therefore to give it more chance of passing into long-term memory) you can **rehearse** that information. For example, when you try to remember a phone number and say it over and over again to yourself, you are using rehearsal to keep the number in your STM. Speaking it out loud or writing it down helps because it is thought that auditory and visual information are more readily stored in STM.

STM does not only process information coming from the sensory memory. It can also act as a location to which information from long-term memory can be brought and used. For this reason it is often called the **working memory**. When doing mental arithmetic, for example, the numbers arrive in the working memory from the sensory memory. By contrast, the instructions about what to do with the numbers arrive from the long-term memory, where they were stored when you first learned to add and subtract.

 Activity

Activity 9.18 investigates which type of information is most successfully stored in STM and the capacity of your STM. **A209ACT18**

 Activity

In **Activity 9.19** you will learn a new skill, using rehearsal to fix it in your memory. **A209ACT19**

Long-term memory (LTM) differs from STM in that there are two types of memories in LTM:

- Memories about facts or events, involving 'knowing that . . .'.

- Memories that we have, but are not consciously fully aware of, involving 'knowing how to . . .'. Memories which allow us to learn motor skills such as writing and walking. Once learnt they are rarely forgotten. You can learn a new motor skill in Activity 9.19.

Information can be stored in long-term memory as meanings (semantics), pictures (visual) and sounds (auditory). Recall from LTM is most effective when we store that information according to its meaning; linking information to previously stored information provides a good cue for recalling it later.

Where are memories stored?

Different types of memory are controlled by different parts of the brain. This is clearly demonstrated by looking at cases where people have lost the use of particular parts of the brain, as the case study below illustrates.

In a patient, HM, who suffered from severe epilepsy, doctors in 1953 removed those areas of the brain that appeared to be causing the problem, and immediately caused amnesia. HM's long-term memories from before the operation were unaffected, but he could no longer form new long-term memories, and found it difficult to remember what he had done just a few minutes before. By contrast, his memory for how to do everyday things was still intact.

Q9.36 Look back at the section of the topic on regions of the brain and find out the parts of the brain that are involved in memory.

How are memories stored in the brain?

In the brain, every neurone connects with many other neurones to make up a complex network. It is by altering the pattern of connections and the strength of synapses that memories can be created.

Eric Kandel, amongst others, studied the molecular biology of learning in the giant sea slug (*Aplysia*, Figure 9.45) to help understand learning in humans. There are no fundamental differences between the nerve cells and synapses of humans and those of lower animals such as the sea slug. However, with only 20 000 neurones, the neurobiology of a sea slug is much simpler than that of a human. They have large accessible neurones (Figure 9.46) so those involved in particular behaviours can be identified. Sea slug behaviour can be modified by learning, and the effects on neurones and synapses studied.

▲ **Figure 9.45** The giant sea slug (*Aplysia californica*). This species may grow to be 30 cm in length and weigh 1 kg.

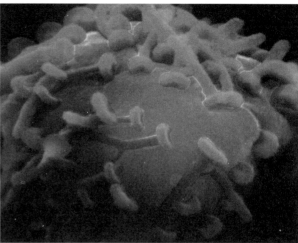

▲ **Figure 9.46** Scanning Electron Micrograph of synapses in the giant sea slug.

Q9.37 a) Why can't the neurones of the sea slug pass impulses by saltatory conduction?

b) How does the sea slug ensure that its neurones can transmit impulses rapidly?

The giant sea slug breaths through a gill located in a cavity on the upper side of its body. Water is expelled through a siphon tube at one end of the cavity. If the siphon is touched, the gill is withdrawn into the cavity (Figure 9.47). This is a protective reflex action similar to removal of a hand from a hot plate.

Because they live in the sea, *Aplysia* are frequently buffeted by the waves. However, they learn not to withdraw their gill every time a wave hits them. They become habituated to the waves. **Habituation** is a type of learning. When you put your socks on you feel them at first, but after a few minutes you no longer notice them. You have become habituated to the feeling of the socks on your feet, even though they are still providing a touch stimulus.

Habituation allows animals to ignore unimportant stimuli so that limited sensory, attention and memory resources can be concentrated on more threatening or rewarding ones.

Kandel stimulated the sea slug's siphon repeatedly with a jet of water. The response gradually faded away, until the gill was not withdrawn any more. The neurones involved in the reflex were identified, and Kandel found that the amount of neurotransmitter crossing the synapse between the sensory and motor neurones decreased with habituation. With repeated stimulations, fewer calcium ions move into the presynaptic neurone when the presynaptic membrane is depolarised by an action potential; fewer neurotransmitter molecules are then released.

Grey matter

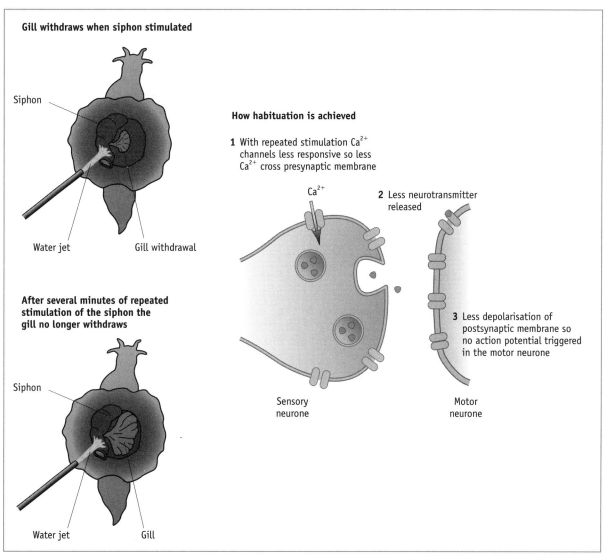

▲ **Figure 9.47** Touching the siphon causes the gill of the sea slug to be withdrawn.

Nice to know: Sensitisation

Sensitisation is the opposite of habituation. It happens when an animal develops an enhanced response to a stimulus. Humans can learn by sensitisation. Imagine you are at home alone late at night, and you hear a loud crash from outside. For a short period after the crash, any small noises (which you previously had not noticed because you were habituated to them) seem very loud and provoke a similar enhanced response. You have become sensitised to these noises.

If a predator attacks, *Aplysia* become sensitised to other changes in their environment and respond strongly to them. Erick Kandel gave an electric shock to the tail before stimulating the siphon again with a jet of water. This provoked an enhanced gill withdrawal response that lasted from several minutes to over an hour. Use Figure 9.48 to help you to understand how sensitisation occurs.

Activity

In **Activity 9.20** you can investigate what is happening at synapses during habituation.
A209ACT20

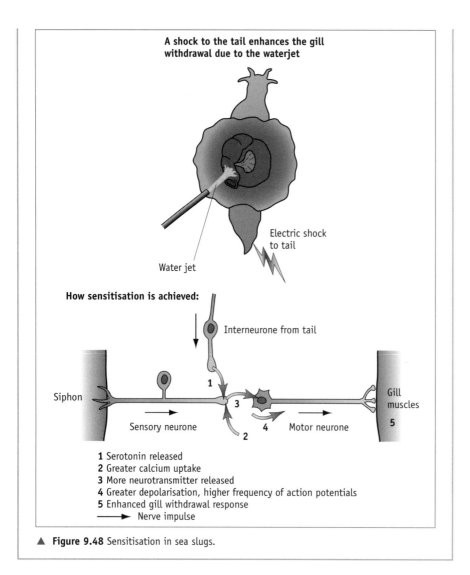

Figure 9.48 Sensitisation in sea slugs.

The changing neurotransmitter release at synapses accounts for memory. Long-term memory storage can also involve an increase in the number of synaptic connections. Repeated use of a synapse leads to creation of additional synapses between the neurones.

The ethics of using animals in medical research

People vary greatly in their views about whether or not it is acceptable to use animals in medical research – or for other purposes such as farming or as pets.

First of all, though their numbers are small, there are those who believe in **animal rights**. People don't any longer consider it morally acceptable to have human slaves so, the argument goes, why is it acceptable to keep animals captive in laboratory cages or on farms? If you believe that humans have certain rights, it is quite difficult to explain why animals have none.

Of course, this doesn't mean that animals such as chimpanzees or dogs would have a right to vote or join a trade union – such rights are as meaningless for them as they are for a three year-old child. But giving laboratory, domestic and farm animals rights would mean that they would have similar rights to such things as food, water, veterinary treatment (equivalent to medical treatment for us) and exercise as we do.

From the point of view of medical research, accepting that animals have rights would mean that we could only use animals that consented to participate in medical experiments, just as we only use humans in medical experiments if they give their consent. In practice, except perhaps for such things as feeding trials, this would bring an end to the use of animals in medical research.

A much more widespread position than believing that animals have rights is believing that humans have a **duty** to ensure that animals are treated well so far as is possible. Here the emphasis is on **animal welfare**. This is pretty much the position in European law. No country in the European Union is allowed to use vertebrates in medical experiments if there are non-animal alternatives. If there are no such alternatives, animals can be used provided strict guidelines are followed (Figure 9.49).

▲ **Figure 9.49** The use of animals in medical research is strictly regulated in the UK and many other countries but remains controversial.

Both the animal rights approach and the animal welfare approach assume that animals can **suffer** and **experience pleasures**. This seems pretty obvious to anyone who has had a pet cat or dog. But what of fish? Can they suffer or experience pleasures? There is genuine scientific and philosophical debate – though the balance is turning in favour of the fish. And what of spiders and insects? Most experts reckon they can't suffer. If you spend your weekends pulling the wings off flies it probably means that you are a most inconsiderate and unpleasant person (or have suffered an abusive childhood); but it probably doesn't increase the amount of animal suffering in the world.

One ethical framework we have introduced in this course is **utilitarianism** – the belief that the right course of action is one that maximises the amount of overall happiness or pleasure in the world (see Topic 2). A utilitarian framework allows certain animals to be used in medical experiments *provided* the overall expected benefits are greater than the overall expected harms. Suppose, for example, to oversimplify greatly, that it takes the lives of 250 000 mice used in medical experiments to find a cure for breast cancer and that 50 000 of these mice are in pain for half their lives. There could still be a utilitarian argument for using the mice.

> **Activity**
>
> In **Activity 9.21** You investigate what people think of using animals for medical research. **A209ACT21**

Q9.38 What would be the utilitarian argument for using 250 000 mice to find a cure for breast cancer?

9.7 Problems with the synapses

The functioning of the nervous system is completely dependent on synapses passing impulses. In everything we do, with every sensation, action and thought, neural pathways are stimulated. The control and coordination of all this activity in the brain relies on synapses. The synapses in turn depend on the neurotransmitters. Unfortunately things can go wrong with adverse consequences for health. Imbalances in the naturally-occurring brain chemicals can cause problems, So can drugs when they cross the blood-brain barrier.

There are about 50 different neurotransmitters in the human central nervous system, and they need to be present in controlled amounts. The function of two examples, dopamine and serotonin, are discussed here, together with the consequences of abnormal levels.

> **Activity**
>
> In **Activity 9.22** you can refresh your memory of how synapses work, and review how they are affected by chemical imbalances. **A209ACT22**

Dopamine and Parkinson's disease

Dopamine is a neurotransmitter secreted by neurones including many located in part of the midbrain. The axons of the neurones in this area extend throughout the frontal cortex, the brain stem and the spinal cord. They are important in the control of muscular movements. Dopamine has also been linked with emotional responses.

In people with Parkinson's disease, these neurones in the basal ganglion die, which means there is almost no dopamine in the brain. The main symptoms of the disease are stiffness of muscles, tremor of the muscles, slowness of movement, poor balance and walking problems. Other problems that may arise include depression and difficulties with speech and breathing. Between 1% and 2% of people in the UK over fifty are affected by Parkinson's disease, although the onset can be well before that age.

Recent developments in drug treatments for Parkinson's have made it much easier for some people to live with the effects of the disease. One approach to treatment aims to slow the loss of dopamine from the brain, with the use of drugs such as Selegiline. This drug inhibits an enzyme called monoamine oxidase, which is responsible for breaking down dopamine in the brain. A second approach is to treat the symptoms with drugs. Dopamine itself cannot be given because it cannot reach the brain from the bloodstream, but L-dopa, a precursor in the manufacture of dopamine, can be given. Once in the brain L-dopa converts into dopamine, increasing the concentration of dopamine and controlling the symptoms of the disease.

Another approach is to give the person dopamine agonists. Dopamine agonists are drugs that activate the dopamine receptor directly. These drugs mimic the role of dopamine in the brain, binding to dopamine receptors at synapses and triggering action potentials. They can be particularly useful in the treatment of Parkinson's disease, since they avoid higher than normal levels of dopamine in the brain. Abnormally high dopamine levels can have unpleasant side-effects. There are also new surgical approaches being trialled, some of which are generating encouraging results.

 Nice to know: Schizophrenia, hyperactivity and the effect of taking cocaine

Excess dopamine in the brain is believed to be a major cause of schizophrenia. Excess dopamine in the brain can be treated with drugs that block the binding of dopamine to its postsynaptic receptor sites. These drugs are usually similar to dopamine in structure, but unable to stimulate the receptors. This reduces the effect of the dopamine in triggering postsynaptic action potentials. A side-effect in patients taking these drugs is to induce the symptoms of Parkinson's disease

Ritalin is a drug used to treat childhood hyperactivity associated with low levels of dopamine. The effect of the drug is to prevent dopamine from being taken back up by the presynaptic membrane; more dopamine is left in the synaptic cleft, to stimulate postsynaptic receptor molecules.

Cocaine also prevents dopamine re-uptake, by binding to proteins which normally transport dopamine. Not only does cocaine bind to the transport proteins in preference to dopamine, it remains bound for much longer than dopamine. As a result, more dopamine remains to stimulate neurones; this causes prolonged feelings of pleasure and excitement. Amphetamines also increase dopamine levels. Again, the result is over-stimulation of these pleasure-pathway nerves in the brain. Long-term amphetamine use can induce schizophrenia-like illness.

 Weblink

Visit the Birmingham Department of Clinical Neuroscience website to see some Parkinson's video clips showing the effect of treatment with L-dopa.

Serotonin and depression

In the brain the groups of neurones which secrete serotonin are situated in the brain stem. Their axons extend into the cortex, the cerebellum and the spinal cord. Neurones which release serotonin target a huge area of the brain (Figure 9.50).

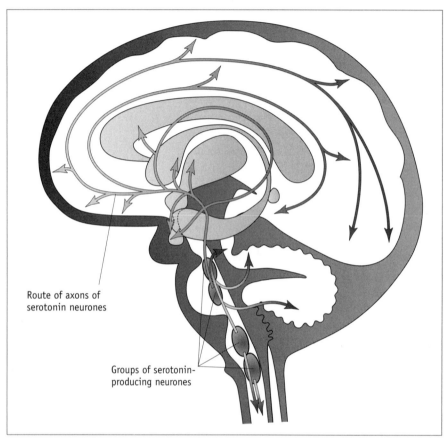

▲ **Figure 9.50** The neurones secreting seretonin are located in the brain stem.

A lack of serotonin has been linked to depression. Clinical depression is a relatively common condition, involving far more than feeling a bit glum. It can last for months or years and can have a profound effect on work and relationships. It is under diagnosed, particularly in children. Depression is associated with feelings of sadness, anxiety and hopelessness. Loss of interest in pleasurable activities and low energy levels are common, as are insomnia, restlessness and thoughts of death.

The causes of depression are not completely understood. There may be a genetic element, since it runs in families. However, for most people it probably has more to do with environment and upbringing than genes. It is probable that a number of factors are involved. Fewer nerve impulses than normal are transmitted around the brain, which may be related to low levels of neurotransmitters being produced. A number of neurotransmitters may have a role to play in depression, including dopamine and noradrenaline; reduced serotonin levels seem to be most commonly involved.

Pathways involving serotonin have a number of abnormalities in people with depression. The molecules needed for serotonin synthesis are often present in low concentrations, but serotonin-binding sites are more numerous than normal, possibly to compensate for the low level of the molecule.

One of the most effective types of drug involved in treating the symptoms of depression inhibits the reuptake of serotonin from synaptic clefts. A drug of this type is a Selective Serotonin Reuptake Inhibitor (SSRI), meaning that it blocks only the uptake of serotonin. One of the more common drugs of this type is **Prozac**, which maintains a higher level of serotinin, and so increases the rate of nerve impulses in serotonin pathways. This has the effect of reducing some of the symptoms of depression.

How can drugs affect synaptic transmission?

Synapses have a number of features that can be disrupted by interference from certain chemicals (Figure 9.51). For example, a chemical with a similar molecular structure to a particular neurotransmitter is likely to bind to the same receptor sites, and perhaps stimulate the postsynaptic neurone. Other chemicals may prevent the release of neurotransmitter, block or open ion channels, or inhibit a breakdown enzyme such as acetylcholinesterase. In all of these cases the normal functioning of the synapse will be disrupted, with consequences that depend on the nerve pathway involved.

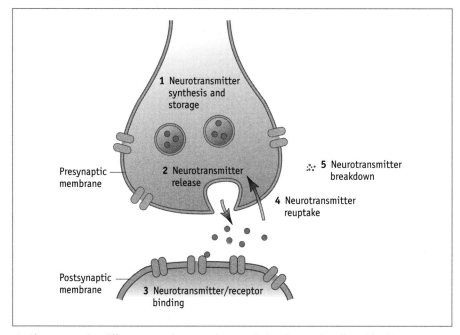

▲ **Figure 9.51** Five different stages in synsaptic transmission that can be affected by drugs. Ecstacy binds to seretonin transport molecules preventing uptake back across the presynaptic membrane.

Topic 9

What effect does ecstasy have on synapses?

The drug known as **ecstasy** affects thinking, mood and memory. It also can cause anxiety and altered perceptions (similar to but not quite the same as hallucinations). The most desirable effect of ecstasy is its ability to provide feelings of emotional warmth and empathy.

Ecstasy is a derivative of amphetamine. Its chemical name is 3,4-methylenedioxy-N-methylamphetamine (**MDMA**). The short-term effects of ecstasy include changes in brain chemistry and behaviour. The long-term effects include changes in brain structure and behaviour.

Ecstasy increases the concentration of serotonin in the synaptic cleft. It does this by binding to molecules in the presynaptic membrane which are responsible for transporting the serotonin back into the cytoplasm. This prevents its removal from the synaptic cleft. The drug may also cause the transporting molecules to work in reverse, further increasing the amount of serotonin outside the cell. These higher levels of serotonin bring about the mood changes seen in users of the drug. It is possible that the ecstasy has a similar effect on molecules that transport dopamine as well.

There are side effects and unpredictable consequences of using the drug. Some people experience clouded thinking, agitation and disturbed behaviour. Also common are sweating, dry mouth (thirst), increased heart rate, fatigue, muscle spasms (especially jaw-clenching) and hyperthermia, as ecstasy can disrupt the ability of the brain to regulate body temperature. Repeated doses or a single high dose of ecstasy can cause hyperthermia, high blood pressure, irregular heart-beat, muscle breakdown and kidney failure. This can be fatal.

It is possible that ecstasy can have an effect on normal brain activity even when the drug is no longer taken. Because the drug has stimulated so much release of serotonin, the cells cannot synthesise enough to meet demand once it has gone, leaving a feeling of depression.

9.8 Genes and the brain

Diseases associated with the nervous system are of major global importance. **Mental illness** is a term encompassing a wide range of diseases and disorders, and is the leading cause of illness and disability in the UK. It accounts for approximately 25% of the Government's total payments on sickness and disability. The sheer misery caused by conditions such as Huntington's disease, Alzheimer's disease, schizophrenia and bipolar disorder (manic depression) means that the search for effective treatments will be one of the major targets for 21st Century biology.

Most disorders of the nervous system do not follow simple Mendelian rules of inheritance (described for monohybrid inheritance in Topic 2 and dihybrid inheritance in Topic 6). One example which does is the condition called Huntington's disease, a degenerative disease of the nervous system. Cells in the basal ganglia and cerebral cortex become damaged, leading to gradual physical and mental changes including involuntary movements, difficulty with speech, mood swings and depression. The phenotypic signs of the disease typically do not appear until the individual is 35–45 years old. The disease is fatal, and death occurs within 15–25 years of onset. The faulty gene responsible occurs on chromosome 4.

Q9.39 a) From the pedigree diagram for a family with Huntington's disease shown in Figure 9.52, decide if the allele for Huntington's disease is dominant or recessive. Give a reason for your answer.

b) Using the letters Hh, state the genotype for each of the 12 people in the diagram.

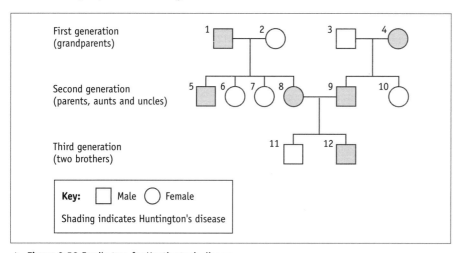

▲ **Figure 9.52** Family tree for Huntington's disease.

Unlike Huntington's disease, most disorders of the nervous system are caused by the interaction of alleles at many loci. When a number of genes are involved in the inheritance of a characteristic, rather than just one, we call the pattern of inheritance **polygenic**. Conditions of the nervous system like Alzheimer's disease and schizophrenia are polygenic as are a wide range of other conditions such as diabetes, coronary heart disease and some cancers.

In many human characteristics that show polygenic inheritance such as height and skin colour there are clearly additive effects of the alleles (see Key biological principle box – Polygenic inheritance). With diseases, several genes may be implicated, but they simply confer a *susceptibility* to the condition, with environmental factors also contributing. Two people who inherit the same susceptibility genes may not both develop the illness; it will depend on environmental factors such as diet, exposure to toxins and stress. Such factors act as triggers to bring about the symptoms of disease. Conditions where several genetic and one or more environmental factors are involved are said to be **multifactorial**.

Twin studies allow determination of the genetic component of a disease. Identical twins, produced by division of the same egg, are genetically identical. The degree of similarity between the twins is a measure of the influence of the genes on that characteristic. 99% of identical twins have the same eye colour, and 95% the same fingerprint ridge count. Where the environment has a greater effect the similarity falls. One study found that if you have Alzheimer's disease and an identical twin sib, your sib has a 40%

chance of having Alzheimer's disease. However, if you have Alzheimer's disease as a non-identical twin, your sib has only a 10% chance of having Alzheimer's disease. This suggests that there is a significant but not inescapable genetic basis to Alzheimer's disease with both genes and the environment influencing the development of the disease.

> **Key biological principle:** Polygenic inheritance
>
> In monohybrid and dihybrid inheritance, each locus is responsible for a different heritable feature. For example, one gene might be coding for the colour of a flower with another gene coding for the shape of the petals. However, some characteristics are controlled by alleles at many loci.
>
> In any introductory course on genetics it is common for eye colour to be used as an example of monohybrid inheritance – a single locus with brown dominant to blue. This is not entirely the case. Eye colour is an example of polygenic inheritance: alleles at several loci control eye colour. Eyes are brown or blue depending on the amount of pigment in the iris. Brown eyes have a great deal of pigment in the iris. This absorbs light, making the eye appear dark. Blue eyes have little pigment, so light reflects off the iris.
>
> Let us say three loci are involved in the inheritance of this characteristic, each with alleles B and b. B adds pigment to the iris and b does not. If all three loci were homozygous for the allele B, the person's genotype would be BB BB BB The additive effect would produce a dark brown iris, whereas bb bb bb would add no pigment to the iris, making it pale blue. A range of possible genotypes and phenotypes are possible between these two extremes, according to how many alleles add brown pigment, as shown below.
>
> | Number of alleles adding brown pigment | Example of genotype | Eye colour |
> |---|---|---|
> | 6 | BB BB BB | very dark brown |
> | 5 | BB BB Bb | dark brown |
> | 4 | BB BB bb | medium brown |
> | 3 | BB Bb bb | light brown |
> | 2 | BB bb bb | deep blue |
> | 1 | Bb bb bb | medium blue |
> | 0 | bb bb bb | pale blue |
>
> The greater the number of loci involved, the greater the number of possible shades.
>
> If a pale-blue-eyed woman has children with a very dark-brown-eyed man they will have light-brown-eyed children as shown below.
>
> | | Mother | Father |
> |---|---|---|
> | Parental phenotypes | pale blue | very dark brown |
> | Parental genotypes | bb bb bb | BB BB BB |
> | Gametes | bbb | BBB |
> | Offspring genotypes | Bb Bb Bb | |
> | Phenotype | Light brown | |

Q9.40 A deep-blue-eyed woman (Bb bb bB) has a child with light brown eyes. Her medium-blue-eyed partner (Bb bb bb) suspects that he is not the father. Complete the Punnett square below and use it to explain to him that he could be the father.

	Mother	Father
Parental phenotypes	deep blue	medium blue
Parental genotypes	Bb bb bB	Bb bb bb
Possible gametes	Bbb, BbB, bbb, bbB	Bbb, bbb

	Bbb	**bbb**

Height, weight and skin pigmentation all involve polygenic inheritance. The combination of polygenic inheritance and environmental effects means that these characteristics show **continuous variation**.

Each allele has a small effect on the characteristic, and the effects of several alleles combine to produce the phenotype of an individual. Suppose that only two genes were involved in the determination of height. The homozygous recessive genotype, aabb, might give a height of 20 cm above a baseline of 140 cm for adult women and 150 cm for adult men. In other words the recessive alleles (a and b) each contribute 5 cm to the height. The dominant alleles, A and B, each contribute 10 cm to the height – so the homozygous dominant AABB would give a height of 40 cm above our baseline. AaBb would add 30 cm to the height.

If two heterozygotes were crossed there would be a range of phenotypes as shown below:

Height above baseline (cm)	20	25	30	35	40
Number of offspring with that height	1	4	6	4	1

If, instead of two loci, there were several, the number of height phenotypes would increase. The more loci, the greater the number of height classes, and the smaller the differences between classes.

Traits such as height and skin colour are not just the result of an individual's genotype. Their environment also has an influence. For example, diet has an effect on a person's height. A poor diet may prevent a person reaching their full height as predicted by their genotype. If you like, their genotype gives their *potential* height. This in combination with their environment determines their *actual* height. The result is that there are no clear height classes, but a continuous variation in height.

> **Activity**
>
> In **Activity 9.23** discover how many genes can affect a single characteristic.
> **A209ACT23**

What can genetics tell us about the structure and activity of the brain?

With the sequencing of the human genome now complete, scientists are working to identify the function of every gene in the human genome. It has been estimated that about 50% per cent of the 30 000 to 40 000 genes in the human genome are expressed in the nervous system. This high level of complexity makes understanding the structure and activity of the brain very challenging.

It is sometimes possible to find genes that are involved in brain disease and to see how gene expression is altered in the diseased tissue. This can provide insights into normal brain development and function – as well as helping in the development of better-targeted drug treatments and other therapies.

Finding the gene

One technique used to find these genes is called **linkage analysis**. This follows the inheritance of DNA markers in affected families. Micro-satellites (see Topic 7) are now being used as markers. The locations of numerous micro-satellites are known, and by using restriction enzymes, PCR and gel electrophoresis it is possible to identify the markers. If a marker is consistently inherited with the disease (Figure 9.53), it suggests that the gene responsible for the disease is located on the chromosome close to the marker. The precise position of the gene can then be located by investigating that section of the DNA in more detail.

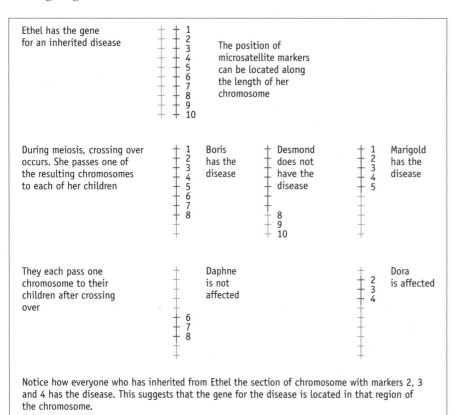

▲ **Figure 9.53** Linkage analysis can be used to locate the position of a gene.

Inheritance of Alzheimer's disease

Six genes have been identified as increasing susceptibility to Alzheimer's disease. In a few families, a genetic fault on chromosome 21 in the APP gene for production of the precursor of a particular protein (known as an amyloid protein) has been implicated in the development of Alzheimer's. Many cases of Alzheimer's have been linked to possession of alleles of a gene known as ApoE. There are three ApoE alleles (ApoE2, ApoE3 and ApoE4), and having two alleles of ApoE4 seems to increase the risk of getting the disease ten-fold. The ApoE gene controls production of a lipoprotein used in the repair of cell membranes in damaged neurones. The ApoE4 allele produces a variant of this, an amyloid protein that is deposited in insoluble plaques in the brain.

Unfortunately how the genes and environmental factors interact to cause Alzheimer's disease is still unclear. It is currently being debated whether many of the dementia conditions are single diseases triggered by a number of genes, or a number of diseases whose symptoms are very similar.

Looking at the expression of genes

Functional genomics has allowed researchers to look at the expression of more than ten thousand genes at any one time, helping us to make comparisons between diseased and normal tissue. If the expression of a gene is changed in a disease cell compared with a normal cell, the gene may well have a role in the disease. Comparing large numbers of genes is done using a microarray. This is a 1 cm^2 grid on a slide. Each of the squares on the grid (and there can be thousands of squares) has the DNA for a specific gene. When genes are expressed, they make mRNA transcripts. All the transcripts produced within a cell are tagged with fluorescent dye. These are then placed on the microarray where they bind to the corresponding genes. In this way it can be determined which genes are being expressed in the cell. Comparisons can be made between healthy and diseased tissue to identify any differences in gene expression.

Scientists can look at gene expression over time and this allows them to look at the development of the brain and how its structure and function change as we age. **Knock-out studies**, where mice are bred with a particular gene switched off, are also used to investigate the role of individual genes in normal brain activity. For example, in studies on the sensory connections from mouse whiskers to the brain, mice have been bred with one gene switched off. This gene was thought to be involved in the organisation of the neurones in the brain into distinct groups, with each group connecting to a single whisker. With the gene switched off the neurones, as expected, failed to organise into groups. Humans don't have whiskers but other knock-out studies are in progress which should tell us about genes that have much the same role in mice and in ourselves, for example the gene involved in Huntington's disease.

We began Topic 9 with Kenge on the African plains mistaking buffalo for insects, and we have ended it by considering knock-out mice, a product of modern biotechnology. Understanding how the brain works requires a

> **Activity**
>
> **Activity 9.24** lets you summarise the methods used to compare the contributions of nature and nurture to brain development discussed throughout the topic.
> **A209ACT24**

remarkable array of approaches which is why this topic contains such diverse things as optical illusions, twin studies, MRI scans, ecstasy use, animal experiments and learning. To a certain extent each of the topics in Salters-Nuffield Advanced Biology has covered a great range of material. Now that you have survived to the end of the course we hope you feel that it has all come together. We are confident that this approach will help you to understand current issues in biology whether you go on to study biology further or not. All the best!

Summary

Having completed Topic 9 you should be able to:

- Understand how the nervous systems of organisms can cause effectors to respond as exemplified by pupil dilation and contraction.

- Explain how a nerve impulse is conducted along an axon.

- Explain how impulses are transmitted across synapses by neurotransmitters (including acetylcholine).

- Describe how the nervous systems of organisms can detect stimuli with reference to rods in the retina of mammals.

- Describe the regions of the human brain that account for our ability to see, think, learn and feel emotions (cerebral hemispheres), thermoregulate (hypothalamus), coordinate movement (cerebellum) and control the heartbeat (medulla oblongata).

- Explain how images produced by MRI and CT scans can be used to investigate brain structure and activity.

- Contrast nervous and hormonal coordination.

- Discuss whether there exists a critical 'window' within which humans must be exposed to particular stimuli if they are to develop their visual capacities to the full.

- Consider the methods used to compare the contributions of nature and nurture to brain development (including evidence from the abilities of newborn babies, animal experiments, studies of individuals with damaged brain areas, twin studies and cross-cultural studies).

- Describe how to investigate visual perception in humans (pattern recognition, optical illusions eg the Müller-Lyer illusion)

- Describe the ways in which animals and humans can learn (perception of stimuli, and memory).

- Describe how to carry out an experiment to characterise learning in humans.

- Describe the role animal models have played in understanding human brain function.

- Discuss the moral and ethical issues related to the use of animals in medical research.

- Explain how imbalances in certain, naturally occurring, brain chemicals (dopamine, serotonin) can contribute to adverse consequences for health (Parkinson's disease, depression) and to the development of new drugs.

- Understand the effects of drugs on synaptic transmission with particular reference to ecstasy (MDMA) and the use of L-dopa in the treatment of Parkinson's disease.

- Explain that some characteristics are controlled by alleles at many loci (polygenic inheritance).

- Describe how genetics is being used to inform the knowledge and understanding of brain structure and activity.

 Review test

Now that you have finished Topic 9, complete the end-of-topic test.
A209RVT01 Congratulations – you have completed the A2!

Answers

Answers to in-text questions for TOPIC 8

Q8.1 B;

Q8.2 a) Strength, flexibility/elasticity; b) Synovial fluid; cartilage;

Q8.3 a) A ligament; B cartilage; C cartilage/pad of cartilage; b) tendon; muscle; synovial membrane; synovial fluid; fibrous capsule;

Q8.4 It would take too long; to move proteins synthesised from mRNA from a single nucleus; to reach the farthermost parts of the cell;

Q8.5 The grey band disappears; because the actin has moved and there is now no point at which there is only myosin on its own;

Q8.6 Better insulation reduces heat loss; so less energy used to maintain core body temperature;

Q8.7 Anaerobic; no oxygen is used;

Q8.8 Oxidation reduction reactions occur as the electron pass along the carrier chain; the final electron acceptor is oxygen;

Q8.9 a) 4; b) 34;

Q8.10 Prevents the cell overheating; allows the controlled release of energy into small useful packets;

Q8.11 Maintain rapid blood flow through muscles to supply oxygen; and remove lactate;

Q8.12 The untrained athlete; larger shaded area on Figure 8.31; they take up and transport oxygen more slowly so take longer to reach maximum oxygen uptake; their period of anaerobic respiration is longer;

Q8.13 a) Glycolysis; and ATP/PC; b) Aerobic;

Q8.14 a) B; b) A; c) C;

Q8.15 Converts fats and proteins to glycogen; for storage in muscles and liver;

Q8.16 a) Cardiac output = SV × HR; = 75 × 70; 5250 cm^3 per minute or 5.25 dm^3 per minute;
b) SV = cardiac output/HR; = 5250/50; 105 cm^3;
c) SV = cardiac output/HR; = 5250/28 = 187.5 cm^3;

Q8.17 a) If pressing on the neck causes increased blood pressure in the carotid artery; blood pressure sensors in the carotid artery would signal to the cardiovascular control centre; which in turn would stimulate the vagus nerve; reducing heart rate and thus pulse measurement;
b) The wrist; or groin;

Q8.18 Anticipatory rise due to the effect of adrenaline on the heart; increases oxygen supply to the muscles in preparation for the activity about to occur;

Q8.19 6 dm^3 min^{-1};

Q8.20 a) 0.65 dm^3; b) 2.7 dm^3; c) tidal volume x rate of breathing; = 0.65 × 18; = 11.7 dm^3 per minute; d) rate of oxygen consumption = volume of oxygen used (dm^3) / time (s); = 0.5 / 20; = 0.025 dm^3 s^{-1} or 25 cm^3 s^{-1};

Q8.21 Rise in pH detected by chemoreceptors; in the medulla, carotid artery and aorta; impulses to ventilation centre; reduction in impulses to the diaphragm and intercostal muscles; no stimulation of the muscles involved in deep breathing;

Q8.22 The depth and rate of breathing increase so there is a greater volume of air inhaled and mixed with the residual air in the lungs; so concentration of oxygen increases; the higher concentration makes the diffusion gradient between the alveolar air and the blood steeper; increasing the speed of gas exchange; as needed given the raised metabolic rate;

Q8.23 Stretch receptors signal the start of movement; allowing ventilation to increase before there is a build up of the waste products of respiration;

Q8.24 Increased concentration of oxygen in the blood; detected by chemoreceptors; slows breathing rate and decreases depth of breathing;

Q8.25 Increased ventilation of the lungs increases oxygen concentration and reduces carbon dioxide concentration; steeper gradient between alveolar air and blood so gas exchange maintained for longer;

Q8.26 a) Aerobic respiration occurs within the mitochondria; large numbers allow slow twitch fibres to have a greater capacity for aerobic respiration;
b) Calcium ions released from the sarcoplasmic reticulum initiate muscle contraction; more sarcoplasmic reticulum allows rapid, repeated contraction of the muscle;
c) Stores oxygen within the cells for use in aerobic respiration;

Answers: Topic 8

   d) Slow twitch via aerobic respiration; fast twitch using anaerobic glycolysis reactions;
   e) Fast twitch; poor supply of oxygen to the fibre; uses anaerobic respiration; rapid build up of lactate;

Q8.27 a) Fast twitch; b) Slow twitch;

Q8.28 a) 37–38 °C; b) Low temperatures lead to reduced metabolic rates as the enzyme-controlled reactions slow; high temperatures increase rate of metabolic reactions initially; but it then declines as the higher temperature denatures the enzymes;

Q8.29 Increases energy heat loss by conduction; and evaporation;

Q8.30 Evaporation from gas exchange surfaces; lowering hairs to increase energy loss by convection; conduction; and radiation;

Q8.31 a) Conduction; b) Some energy is transferred to the body cells as a waste product of normal metabolism; shivering increases resting metabolism 3- to 5-fold; transferring additional energy to the body cells; nerve impulses to the arterioles in the skin cause vasoconstriction; resulting in restricted blood flow through the skin; this reduces energy loss by radiation, conduction and convection; hair raising is pretty ineffectual in humans, particularly for cross-Channel swimmers who coat themselves in protective grease (petroleum jelly);

Q8.32 Reduced risk of upper respiratory tract infections with moderate volume or moderate intensity exercise; increased risk of infection with large amounts of or high intensity exercise;

Q8.33 a) B cells;
   b) Inflammation; phagocytosis; anti-microbial proteins;
   c) Killer T cells will not be activated by the cytokines;

Q8.34 Moderate exercise increases the number of natural killer cells; intense exercise reduces the number and activity of natural killer cells, phagocytes, lymphocytes and helper T cells;

Q8.35 Increase the number of red blood cells and hence the amount of haemoglobin; improving blood's oxygen carrying capacity; enhancing oxygen delivery to muscle tissue and hence improving aerobic capacity;

Q8.36 Blood clots in the arteries or veins;

Q8.37 Sprint events rely on anaerobic respiration so performance is not dependent on the athlete's aerobic capacity;

Q8.38 To help determine if athletes are taking EPO as a performance-enhancing drug;

Q8.39 Lipid-based;

Answers to in-text questions for TOPIC 9

Q9.1 They produce rapid responses; important for protection and survival;

Q9.2 Other neurones must be involved; which are under conscious control;

Q9.3 The one on the left;

Q9.4 Radial;

Q9.5 a) Rods and cones cells in the retina; b) Optic nerve; c) Brain; d) Iris muscle;

Q9.6 Prevent damage to the retina with high intensity light; in dim light it ensures maximum light reaches the retina;

Q9.7 Protect the eye from sudden flashes of bright light;

Q9.8 Three;

Q9.9 The channel's opening is due to changes in voltage;

Q9.10 No (unless ATP was added); the polarisation of the membrane is maintained by the concentration gradients achieved by the action of energy-requiring sodium-potassium pumps; membrane integrity is lost;

Q9.11 A new action potential will only be generated at the leading edge of the previous one; because the membrane behind it will be recovering/incapable of transmitting an impulse;

Q9.12 Might expect photoreceptors to be on the surface of the retina, but they are a deeper layer/light has to travel through the other layers including blood vessels before reaching the photoreceptors;

Q9.13 a) i) Active transport; ii) Diffusion;
b) One that lets any positive ions through, such as Na^+ and Ca^{2+};
c) Because sodium ions are being actively transported out; and their re-entry through ion channels is prevented;

Q9.14 Any orange-coloured food, for example oranges, carrots, chillies;

Q9.15 The region of the brain concerned with vision processing is the occipital lobe which sits at the back of the cortex and is closest to the back of the head (thus a blow to this area would cause a disturbance in vision);

Q9.16 Frontal lobe; parietal lobe; motor cortex; cerebellum;

Q9.17 Parietal lobe/basal ganglia;

Answers: Topic 9

Q9.18 There are many conditions that can cause brain damage, including lack of oxygen, carbon monoxide poisoning, toxic chemical exposures, infectious diseases, tumours, strokes and genetic conditions;;;

Q9.19 Cerebellum;

Q9.20 a) i) fMRI; ii) MRI;
b) The posterior hippocampus is involved in remembering detailed mental maps;

Q9.21 Visual stimulation with both light and patterns;

Q9.22 Because kittens are born blind, early deprivation (under three weeks) would have no effect; by three months connections to the brain have been made, and deprivation has no effect / the critical period has ended; the critical period is at about four weeks of age so lack of stimulation from the kitten's environment at this time severely affects visual development;

Q9.23 For some birds their song seems to be innate, due to nature and not nurture; for others both nature and nurture seem to be required;

Q9.24 1;5;3;2;4 or 1;5;3;4;2;

Q9.25 strawberry; pencil;

Q9.26 a) Bart Simpson; b) Marilyn Monroe; c) Michael Reiss (Director of the SNAB project); Anyone who could not be recognised is probably due to never having seen the individual before;

Q9.27 Synapses from the optic nerve axons to the visual cortex will have been weakened or eliminated; the visual cortex may only receive sensory information from one eye; so the binocular cells will not have the second view to compare this with, making stereoscopic vision difficult or impossible;

Q9.28 Familiarity with the relative sizes of the elephant and antelope; overlap, one hill partly hiding another tells us it is closer;

Q9.29 The fact that there is a correlation between two variables does not necessarily mean that there is a causal relationship;

Q9.30 Some apparent cross-cultural differences in perception may occur because people cannot report perceptual differences; there may, for instance, be a language difficulty; researchers from Western societies chose the tests; there has been a strong emphasis on two-dimensional visual illusions which may favour subjects from the carpentered world;

Q9.31 She will show symptoms of distress or fear, and may refuse or cry;

Answers: Topic 9

Q9.32 a) The baby has not experienced this before so cannot have learnt it;
b) Some perceptual development has certainly taken place since birth; the experiment requires the baby to crawl and this isn't possible for several months;

Q9.33 The fact that these animals had had little time for learning suggests the behaviour is innate;

Q9.34 Kenge was not used to vast open spaces and would have had little opportunity to look far into the distance; with no experience of seeing distant objects he would have little experience of certain environmental depth cues such as *size constancy* – the tendency to see the same object as always being the same size, however far away it is; as a result he saw the buffalo simply as very small animals;

Q9.35 a) neutral stimulus – sound; unconditioned stimulus – receiving food; conditioned response – salivation;
b) neutral stimulus – sound; unconditioned stimulus – puff of air; conditioned response – blinking;

Q9.36 Parietal lobe; hippocampus;

Q9.37 a) Neurones not myelinated;
b) Large diameter axons;

Q9.38 Breast cancer is common in women; over time hundreds of thousands of women die from it; each woman may suffer more than each mouse; friends and relatives of a woman who dies young also suffer emotion pain;

Q9.39 a) Allele dominant; individuals 8 and 9 who both have the condition produce an offspring who does not; individuals 8 and 9 must be heterozygous;
b) 1 Hh; 2 hh; 3 hh; 4 Hh; 5 Hh; 6 hh; 7hh; 8 Hh; 9 Hh; 10 hh; 11 hh; 12 Hh;

Q9.40

	Bbb	bbb
Bbb	BB bb bb (2)	Bb bb bb (1)
BbB	BB bb Bb (3)	Bb bb Bb (2)
bbb	Bb bb bb (1)	bb bb bb (0)
bbB	Bb bb Bb (2)	bb bb Bb (1)

The number of alleles adding pigment is shown in brackets. The child with 3 alleles will have light brown eyes. There is a 1 in 8 chance of the couple having a child with brown eyes;;;;